Self-Organization of Hot Plasmas

Yu.N. Dnestrovskij

Self-Organization of Hot Plasmas

The Canonical Profile Transport Model

Yu.N. Dnestrovskij
Kurchatov Institute
Moscow
Russia

Enhanced translation from the Russian Edition:
Самоорганизация горячей плазмы (Samoorganizatsija gorjachej plasmi)
by Yu.N. Dnestrovskij

ISBN 978-3-319-06801-5 ISBN 978-3-319-06802-2 (eBook)
DOI 10.1007/978-3-319-06802-2
Springer Cham Heidelberg New York Dordrecht London

Library of Congress Control Number: 2014939522

Printed on acid-free paper

Springer is part of Springer Science+Business Media (www.springer.com)

Preface to the English Edition

This book is devoted to the problem of confinement of energy and particles in tokamak plasmas. Although the first tokamaks were constructed more than half a century ago, large efforts will still be needed before a detailed description of transport in these devices will be a reality. So far, high hopes for a description of transport in tokamaks using huge multidimensional gyro-kinetic codes did not materialize. Growing attention is now focusing on the idea of self-organization in plasmas. The fusion community rejected a long ago the fundamental concept of a local connection between transport coefficients and plasma parameters. The idea that the fundamental nature of transport in fusion plasmas is dual, begins now to gain momentum: profiles of pressure and temperature are determined in main by the magnetic configuration and in a lesser extent by the energy and particle fluxes.

The English edition of this book is very close to the Russian edition. In the first five Chapters only editorial corrections were made. The only addition is Sect. 6.9 in Chap. 6, devoted to the analysis of temperature and pressure pedestals of JET ELMy H-mode discharges. The comparison of the experimental pedestal values with the canonical profile pedestals allowed us to establish simple relations to calculate temperature and density pedestals in general transport calculations. These relations are true for tokamaks with moderate aspect ratio $A = R/a \sim 2.5–4$.

I want to express my sincere appreciation to my colleagues S.E. Lysenko and I.S. Marchenko for their invaluable help with the translation. Without them it would have been a nearly insurmountable task to prepare the English version. First contacts with the Springer Publishing House were made by J. Ongena. He also made many suggestions to the translated text of this book. I want to express him my deep gratitude. Maria Bellantone and Mieke van der Fluit benevolently helped me during a long and tensional work with Springer. I bring them my hearty thanks.

I presented the Russian Edition of this book to many colleagues in the West and East. It is my hope that this English Edition will help them to provide an easier access to the ideas developed in the Russian version.

March 2014 Yu.N. Dnestrovskij
Moscow, Russia

Preface to the Russian Edition

Work on controlled thermonuclear fusion began in our country on the initiative of I.V. Kurchatov in the early 1950s. But the initial hope for a rapid realization of the final goal could not be fulfilled. The problem turned out to be much more difficult than originally thought, both in getting an understanding of the physical processes in the hot plasma and in surmounting the technological challenges inherent in the practical realization of heating and confining hot plasmas. The difficulties soon led to a united effort of scientists and engineers worldwide, culminating in the ITER project, a unique example of international cooperation. Confidence now grows that the first tokamak reactor could be built in the middle of the third decade of the twenty-first century.

However, the understanding of the physical processes underlying the energy and particles transport in a tokamak plasma is far from simple. Transport is mainly determined by turbulent processes in the plasma, and binary collisions between the particles play a secondary role. In principle one should be able to describe turbulent transport by solving numerically multi-dimensional kinetic equations using so-called gyro-kinetic codes. However, after more than two decades of huge efforts using the fastest computers in the world, one is far from a workable solution along this road. The main difficulty is the large difference in various characteristic times (6–7 orders of magnitude) and characteristic scales (4–5 orders of magnitude) inherent in the problem. In addition the experimentally observed large-scale self-organization of plasma (strong tendency to conserve pressure and temperature profiles), can still not be reproduced using gyro-kinetic codes.

The lack of a universal transport model forced scientists to look in other directions for the analysis of existing experiments and the extrapolation to future devices. A 'brute force' solution is the use of purely empirical scalings based on the analysis of a large database of experimental data obtained from a various fusion devices around the world. In this way work has continued over the last 25 years to determine scalings for the energy confinement time and other characteristic parameters in various plasma regimes. One of these scalings is the basis for the design of ITER.

It is clear that such scalings are only a first step towards a description of the transport processes in a thermonuclear plasma. The next step is to create one-dimensional (1D) models that are based on theoretical principles. The presentation and justification of such a model is the purpose of this book.

The earliest 1D models to describe the transport of energy and particles in fusion plasmas date back to the late 1960s. Transport coefficients were derived from experimental data and depended on local plasma parameters. This continued until the middle of 1980s, when it became clear that effects of self-organization play an important role in transport and must be incorporated into the model. First proposals for such 1D models were formulated by different authors, including ourselves, at the EPS Conference on Plasma Physics in Dubrovnik in 1988. In what follows, such models are called "critical gradient models". In our model the critical gradient is closely linked to the self-organization of the plasma, and thus it is different from other similar models. We assume that the critical gradient is determined by the minimum of the magnetic energy of the toroidal plasma current with condition of the conservation of total current and total poloidal magnetic flux. The other models use usually the border of the instability of some drift waves as a condition for the critical gradient. The book first discusses the underlying theoretical hypotheses, based on a large number of experimental observations, that allow us to describe the phenomenon of self-organization and then the transport model is proposed using the concept of critical gradients. The transport coefficients in this model are determined by comparing the calculations with experimental results.

This book is organized as follows. In the Introduction basic concepts concerning self-organization of plasmas are discussed, and illustrated with examples from experiments. In the second chapter a variational formulation is presented for the so-called 'canonical' temperature and pressure profiles. Experimental profiles evolve towards canonical profiles and fluxes of heat and particles are determined by the difference (in some metric) between the experimental and canonical profiles. Since the plasma is an open system of particles and energy, and fluxes are always present in plasmas, the experimental profiles will never coincide with the canonical ones, but approximate them to a larger or lesser extent.

The third chapter extends the ideas on plasma self-organization to stellarators. Such a description for stellarators is more complex than for tokamaks, since the absence of axial symmetry in the plasma leads to large neoclassical fluxes that can compete with turbulent fluxes. We present experimental observations on the conservation of the pressure profile in stellarators and propose a variational formulation for canonical profiles in such devices.

The fourth chapter discusses scaling laws for the plasma energy confinement time. We discuss multi-machine scaling laws derived from a large multi-machine database (in particular ITER scaling) and one-machine scalings. The Taylor—Connor theory of invariants is presented, together with its conditions on the structure of scalings and transport coefficients.

The fifth chapter presents the linear version of the canonical profiles transport model. We discuss its application to the ohmic regime and L-mode regime in tokamak discharges. The plasma density, the electron and ion temperatures, the velocity of toroidal rotation and the toroidal current density are the required variables in this model. It can also be applied to the H-mode if the estimates for the pedestal values are known.

Finally, in the sixth chapter we discuss the non-linear version of the transport model, for the description of improved confinement regimes with external and internal transport barriers. H-mode plasmas with an external barrier at the plasma edge can be satisfactorily simulated. The same cannot yet be said about the description of plasma regimes with internal barriers, as the formation of such a barrier is linked to the position of resonant surfaces, and thus their position has to be determined by the discharge scenario, including the time evolution of the current profile. The chapter discusses also the not fully resolved issue of the impact of the toroidal rotation velocity on plasma confinement. Although we know experimentally that this effect is small, still more work is needed for a good quantitative description.

I started to work on transport models in the 1960s on the initiative of Academician L. A. Artsimovich together with D. P. Kostomarov, to whom I am deeply grateful. For several decades we worked together, discussing daily difficulties and progress. This close collaboration was extremely helpful to develop new ideas and improve the understanding of the main features of transport in tokamaks, resulting in the 1980s in several books on the modeling of fusion plasmas. The present book could not have been written without the lessons learned over many years working with him.

The basic theoretical ideas were developed by Academician B. B. Kadomtsev. As early as 1986–1987 he proposed to use a variational formulation as the basis for a self-consistent description of the plasma profiles observed in experiment. The main experimental background to this book is from the papers by B. Coppi (1980) and Y. V. Esipchuk and K. A. Razumova (1986). Fruitful and intense discussions with Ksenia Alexandrovna helped to crystallize many of the concepts presented in the subsequent chapters.

In preparing the book for publication I especially thank S. E. Lysenko for his invaluable assistance and hard work in the general plan and design of this book. Special thanks also go to my pupils and colleagues: A. Yu. Dnestrovskij, V. F. Andreev, A. V. Melnikov, A. V. Danilov, K. N. Tarasyan and S. V. Cherkasov. An especial gratitude to V. S. Mukhovatov for his patience and support in discussing the main ideas of this book, still not fully accepted by the scientific community. Thanks also to A. M. Stefanovskij for intensive discussions on the nature of variational problems.

For over 15 years our group is collaborating with scientists of the Culham Centre for Fusion Energy (CCFE), Culham, UK. Many of the ideas of this book were shaped during discussions with John Connor, Antony Field, Tim Hender, Colin Roach, Martin Valovich, Mikhail Gryaznevich, Irina Voitsekhovitch and Michele Romanelli. To all of them I am deeply grateful.

September 2012 Yu.N. Dnestrovskij
Moscow, Russia

Contents

1 Introduction ... 1
References ... 8

2 Variational Principles for Canonical Profiles in a Tokamak ... 11
2.1 The Principle of Total Energy Minimum by Hsu and Chu ... 11
2.2 The Principle of Minimum of the Plasma Current Magnetic Energy for a Circular Plasma Cylinder (Kadomtsev) ... 14
2.2.1 The natural but contradictive statement of the variational problem ... 14
2.2.2 The Adjusted Statement of the Variational Problem ... 19
2.3 Canonical Profiles for Toroidal Plasma with Arbitrary Cross-Section ... 25
2.4 Examples ... 29
2.5 Canonical Profiles of the Toroidal Rotation ... 33
References ... 39

3 A Possible Approach to the Canonical Profiles in Stellarators ... 41
3.1 Original Equations ... 41
3.2 Variation of Energy ... 43
3.3 Application of Formulas to Stellarators ... 44
3.4 Canonical Pressure Profiles ... 46
3.5 Approximate Solution of Equilibrium Equations ... 46
References ... 49

4 Theoretical Limitations for Scaling Laws and Transport Coefficients ... 51
4.1 Plasma Energy Confinement Time and Scaling Laws ... 51
4.2 Interplays Between Parameters Describing the Plasma State ... 52
4.3 One-Machine Scaling ... 55
4.4 Multi-Machine Scaling ... 57

4.5 Examples 59
4.5.1 Multi-Machine Scaling ITER 59
4.5.2 Experimental One-Machine Scaling Laws 60
4.5.3 A Comparison of Scalings 62
4.6 Connor-Taylor Theory of Invariants 63
4.6.1 Transformation of the Complete System of Equations for the Plasma 63
4.6.2 Approximation of Quasi-Neutral Plasma 64
4.6.3 Approximation of a Collisionless Plasma 67
4.6.4 Restrictions on the Transport Coefficients 69
References 70

5 Linear Version of the Canonical Profiles Transport Model (CPTM) 71
5.1 Transport Equations 71
5.2 Heat Flux, Particle Flux and Flux of Toroidal Momentum in Ohmic and L mode Plasmas 72
5.3 Definition of the Transport Coefficients 77
5.4 The Structure of the Full Transport Model 79
5.5 Stiffness of the Heat Diffusion Equation 79
5.6 Examples 81
References 89

6 Nonlinear Version of the Canonical Profiles Transport Model (CPTM) for Improved Confinement Regimes 91
6.1 Plasma Regimes with Improved Confinement 91
6.2 Heat and Particles Fluxes in Improved Confinement Regimes 92
6.3 Approximate Analytical Criterion for the L–H transition 95
6.4 Estimates of Transport Barrier Parameters in H-Mode 97
6.5 Transport Coefficients in the Nonlinear Model 100
6.6 General Remarks on the Second Critical Gradient 104
6.7 Examples 111
6.8 Remarks on the Stiffness of the Ion Temperature Profile 120
6.8.1 Radial Dependence of the Stiffness 120
6.8.2 Dependence of the Stiffness on the Toroidal Rotation Velocity 124
6.9 Approximation for the Pedestal Values Based on the Experimental Data 127
6.9.1 General Expressions 127
6.9.2 Estimates for the Normalized Temperature Based on JET Experimental Data 128
References 131

Concluding Remarks 133

Chapter 1
Introduction

Abstract The basic concepts governing plasma self-organization, i.e. the conservation of the profile shape of specific plasma parameters (pressure, temperature, toroidal rotation velocity) under the influence of external sources of heat, particles and torque are discussed. Examples of the self-organization in tokamaks with circular and elongated cross-sections are shown. The possibility of plasma self-organization in stellarators is considered. A brief review is given of the mathematical models describing the canonical profiles of plasma parameters.

The remarkable property of a tokamak plasma that tries to maintain the shape of the spatial profiles of temperature and pressure at external influences, has been known for a long time. B. Coppi [1] was probably the first who clearly formulated in the early 80's the idea of optimal or "canonical" profiles for plasma parameters. A detailed study of the experimental profiles from different devices with circular plasma cross-section, carried out in that years by Yu.V. Esipchuk and K.A. Razumova [2] confirmed the ideas formulated in [1].

The property of the plasma to maintain the profiles of some parameters is called "self-consistency" or simply "self-organization of the plasma." Of course, the profile shape is not maintained in an 'absolute' way, and as always, "the devil is hidden in the details." We shall see that these details define the plasma transport. Omitting details, the self-organization of the plasma can be summarized as follows: the radial shape of several plasma parameters in many different plasma confinement modes tends to be close to a selected profile shape, which we call "canonical". The transition from an arbitrary initial profile shape to the canonical one will be called relaxation. The meaning of self-consistency is reflected in the property of the plasma to "remember" the canonical profile and to "direct" the relaxation process in that direction, thereby changing the transport coefficients.

Thus, to describe the self-consistency of a tokamak plasma one needs to solve two problems: first, to build a mathematical model for the canonical profiles, and second, to describe the relaxation of the plasma profiles to the canonical ones by constructing a suitable transport model and a set of transport equations. The first chapters of this work are devoted to various approaches for the construction of the canonical profiles. In subsequent chapters the transport model is derived, and both the relaxation process and the derivation of the steady state experimental profiles from the canonical ones are discussed.

Yu.N. Dnestrovskij, *Self-Organization of Hot Plasmas,* DOI 10.1007/978-3-319-06802-2_1,

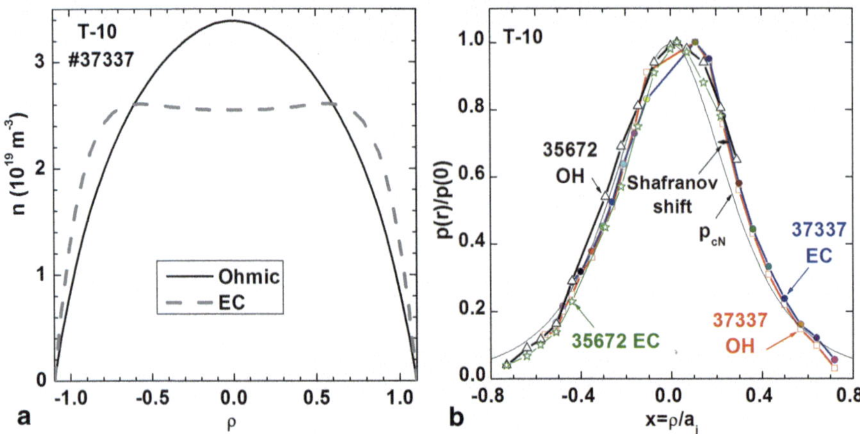

Fig. 1.1 **a** Plasma density profiles in T-10 device in the Ohmic regime (*OH*) and in the regime with ECR heating (*EC*). **b** Electron pressure profiles normalized in the plasma centre for T-10 discharges with different currents in OH and EC regimes. The normalized radius $x=r/a_j$ (self-similar variable) is used over the horizontal direction. Here $a_j=(IR_0/B_0)^{1/2}$, I is the plasma current in MA, R_0 is the plasma major radius in m, B_0 is the toroidal magnetic field in T [4]

With the idea of plasma self-organization, we challenge many conventional ideas about the transport of particles, energy and angular momentum [3]. Fluxes of particles, energy and momentum can be nonlinear functions of gradients. Transport coefficients may be non-local functions of the parameters of the plasma and may depend on the source distribution. At first sight, the self-organization of the plasma as proposed here raises more questions than it answers. Which parameters of the plasma are subject to self-organization? What is the form of the canonical profiles? Which plasma profiles are more stiff? What determines the rate of relaxation? All of these issues are discussed and clarified in this book.

In order for the reader to understand the relationship between plasma parameters within the context of self-organization, some examples may be instructive. Figure 1.1a, shows typical profiles of the plasma density in the T-10 tokamak for a discharge with a pure Ohmic heating (OH) phase and a phase with additionally central microwave heating of the electrons (EC) [3]. In what follows, we define ρ as the radial coordinate, with $\rho>0$ and $\rho<0$ for the external and internal surfaces of the torus, respectively. In the OH phase, the density profile is monotonous and rather broad. When switching on the microwave power the particles are pushed out of the heating zone and the density profile becomes non-monotonic. In the literature, this process is called "pump-out effect". We describe such a density profile as "soft", i.e. it easily gives in to outside influences and can vary considerably.

Figure 1.1b, shows the profiles of the electron pressure for the T-10 shots #37337 and #35672, normalized to the central point $\rho=0$

$$\rho_{eN}=\frac{n_e(\rho)T_e(\rho)}{n_e(0)T_e(0)} \tag{1.1}$$

in the OH and EC modes. Here $n_e(\rho)$ and $T_e(\rho)$ are the electron density and electron temperature. The temperature profiles were taken at several time instances just before the breakdown in the sawtooth oscillations. For the shots selected here the difference in plasma current I is about a factor of 1.4. Therefore, if we select the radial coordinate in the figure to be the normalized value $x = \rho / a_j$, where $a_j = (IR_0 / B_0)^{1/2}$, with R_0 the major plasma radius, and B_0 the toroidal magnetic field at the plasma axis, it can be seen that in these coordinates all 4 curves coincide within experimental accuracy. This means that changes in temperature and density profiles during microwave heating cancel each other and as a result, the pressure profile is almost unchanged. The constancy of pressure profiles in Fig. 1.1b confirms the independence of the shape of the profile on the type of heating power (Ohmic or microwave heating) and points to the existence of self-similar radial variable x, at which the pressure profiles are similar at different values for the plasma current and toroidal magnetic field.

It seems at first glance, that if the density profile has changed considerably, as shown in Fig. 1.1, then, as a consequence the temperature profile (by (1.1)) should also change dramatically, in order to maintain the pressure profile. However, this is not the case. To show this, we estimate the change in the temperature profiles, for the examples discussed above. Omitting for brevity the index 'e', we have

$$p = nT, \quad \frac{p'}{p} = \frac{n'}{n} + \frac{T'}{T}. \tag{1.2}$$

Here $p' = dp / d\rho$. Since the pressure profile is preserved during the ECR heating, therefore $p' / p = \text{const}$ in this process. In the steady-state phase of shot #37337 $|T'/T| \sim 2|n'/n|$. If the value of n'/n varies by 10%, due to (1.2), the value of T'/T should vary by only 5%. In reality, the pressure profile itself (the value of p'/p) also may change a couple of percents, imperceptible in Fig. 1.1b, therefore, to preserve Eq. (1.2), a change in T'/T of only 2–3% is sufficient. This is what we mean by the statement that the temperature profile is stiffer than the density profile. A detailed comparison of the profile stiffness of the temperature and pressure will be carried out in the following chapters.

Another example is provided by two Ohmic MAST shots with the same plasma current, but at different densities [4]. Normalized at r=1.05 m, the density profiles at time t=0.15 s for shots #11446 and #11447 are shown in Fig. 1.2a. Here r is the distance from the main axis of the torus. The electron temperature profiles were obtained by measuring the Thomson scattering of the laser beam in 300 space points. In shot #11447 the high-density plasma temperature is lower, so the current rapidly penetrates into the plasma core. As sawtooth oscillations with mixing in the center were present, the density profile is flatter than in shot #11446. Figure 1.2b, shows the electron pressure profiles for both shots, normalized to the same point, r=1.05 m. It can be seen that outside the sawtooth region the pressure profiles are virtually identical.

What determines the type of canonical profiles? It is frequently assumed that it is determined by the extremum of some functional (magnetic [5–7] or total plasma

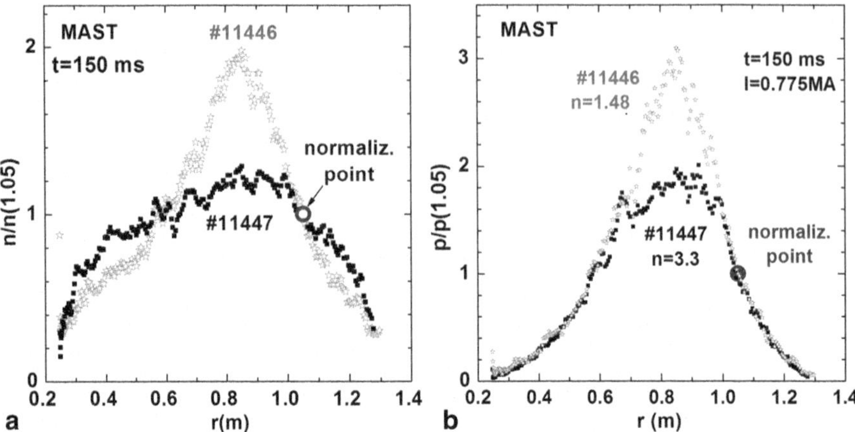

Fig. 1.2 **a** Plasma density profiles in MAST device normalized in the point $r=1.05$ m in Ohmic regime. In the discharge #11447 plasma density is 2.2 times higher than in the discharge #11446. **b** The profiles of the normalized electronic pressure $p_e=nT_e$ for the same discharges. In the discharge #11447 the sawtooth oscillations mix the pressure in the plasma core [4]

energy [8], the growth rate of the magnetic entropy [9, 10], energy dissipation rate [11–14] etc.). In other papers, the canonical profiles are found from the instability criterion of different drift modes: the ion temperature gradient mode (ITG), the electron temperature gradient mode (ETG), the trapped electrons mode (TEM) [15–17]. The form of the functional as well as the type of drift modes, are not absolutely defined. But even after selecting a certain functional, one cannot be sure to find the final form for the canonical profile. Typically, the number of unknown functions exceeds the number of Euler equations, and one needs to make additional assumptions to allow a unique determination of the canonical profiles.

The physical interpretation of the phenomenon of self-organization of the plasma is also quite uncertain. It is impossible to identify a single physical mechanism responsible for the canonical form of the profile and the rate of relaxation. It is usually assumed that the turbulence generated by the various instabilities that determines anomalous transport of plasma parameters. For over 20 years gigantic gyrokinetic codes describing turbulence try to address the problem of a self-consistent description of plasma transport. In a recent paper [18], it is shown using such a code that the evolution of the plasma turbulence leads to a transport pattern that can be described as follows (currently only for the ion component of the plasma): during the relaxation the ion temperature profile is slightly changed, and is followed by an avalanche-like process that resets the plasma energy. Such a pattern can serve as a prototype for the self-organization effect of the plasma. However, there is still a large step to make before applying these results to the description of real experiments.

In this book we develop the ideas proposed by B.B. Kadomtsev [6] and by D. Biskamp [7], which state that the canonical profiles are determined by the minimum of the magnetic energy functional of the toroidal plasma current with the additional condition that the current and the poloidal magnetic flux are conserved. In this formulation, the two functions describing the poloidal magnetic field and the plasma pressure are unknown. To close the problem, we assume that the canonical profiles of current and pressure coincide. In this case, a realistic boundary condition allows us to determine a unique solution for the canonical profiles. An easy variational formulation of the problem allows us to apply the results obtained to formulate a relaxation model for the plasma transport.

In the work by Hsu and Chu [8] it is assumed that the canonical profiles are determined by the minimum of the total plasma energy functional including the energy of the toroidal magnetic field. At first glance, it appears that this approach is more general and more promising than the approach of [6], [7]. However, they were not able to build a model for the relaxation of the plasma profiles and consequently in the literature this work has not been continued. Based on the ideas of [8], it can be shown that there are arguments supporting the coincidence of the canonical profiles for current and pressure. The second chapter of this book is devoted to discussing and summarizing the results of works mentioned above.

The plasma toroidal rotation velocity υ_t for medium-sized tokamaks can be divided into 2 types:

1. Spontaneous rotation with an experimental rotation velocity υ_t usually in the range

$$\upsilon_t < 20 - 40\text{km/s}. \tag{1.3}$$

2. Driven rotation with much greater velocities in the range:

$$\upsilon_t < 200 - 400\text{km/s}. \tag{1.4}$$

The origin of spontaneous rotation is discussed extensively in the literature [19–22]. It is believed that the source of this rotation is the radial electric field E_r, arising due to various reasons, and the poloidal magnetic field B_θ associated with the plasma current. The drift velocity is then determined by the relation

$$\upsilon_{dr} = \upsilon_t = cE_r / B_\theta \tag{1.5}$$

Driven rotation is determined by the external toroidal torque t_L, introduced into the plasma. The most common source is the beam of hot neutrals injected into the plasma. In this book, we discuss only driven rotation. Using the procedure of [8], one can construct a canonical profile for the angular velocity of rotation, which is coincident with the canonical profile of the plasma density.

We see that the profiles of temperature, pressure, and even the angular velocity of rotation in tokamaks are self-consistent. However, for stellarators this issue is still not resolved. There are studies that suggest that in stellarators the temperature profiles do not possess the property of self-consistency [23] and similarly for the

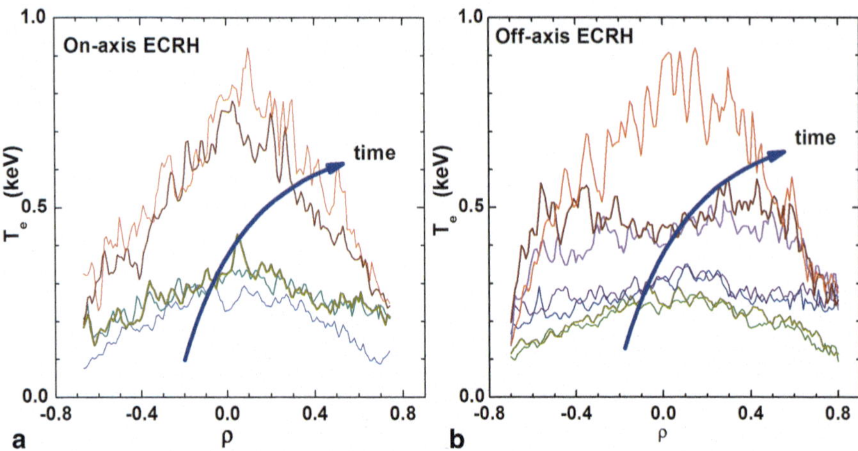

Fig. 1.3 The electron temperature profiles at the on-axis (**a**), and off-axis (**b**) ECR heating in the stellarator TJ-II [25]

pressure profiles. The majority of the scientific community believes that plasma transport is mainly determined by neoclassical effects in the stellarator. However, the discovery of the H-mode and internal transport barriers in a stellarator does not fit into a purely neoclassical transport paradigm. In addition, it was shown recently in [24] that for several stellarators the pressure profiles are self-consistent. This fact contradicts the neoclassical picture. Apparently, anomalous transport due to plasma turbulence still plays an important role in stellarators. But, if indeed so, the assertion that there is no self-consistency for the temperature profiles in stellarators may also be subject to revision. This requires more work on the analysis of the temperature profiles in order to solve this problem once and for all.

To illustrate the above we give, as an example, the experimental data obtained in the stellarator TJ-II [25]. Figure 1.3a, b shows the electron temperature profiles for on-axis (a) and off-axis (b) EC resonance heating. In the second case the temperature profiles are flat and even hollow. The corresponding normalized electron pressure profiles are shown in Fig. 1.4. It is seen that all of the profiles are similar and possible deviations do not exceed the experimental errors.

In Chap. 3 dedicated to the analysis of the existence of canonical pressure profiles in stellarators, we use a variational procedure proposed in [8]. In this way, in addition to solving the general problem in the approximation of a large aspect ratio $R_0/a >> 1$ and circular cross section of plasma, it is possible to construct an analytical formula for the canonical pressure profile, convenient to analyze experiments. However, the unresolved issue of self-consistency of the temperature profiles does not yet allow us to build a model for energy and particle transport in a stellarator based on canonical profiles.

The transport model that describes the relaxation of the plasma parameters to the canonical profiles in tokamaks, was developed in [26–30]. In this model, the

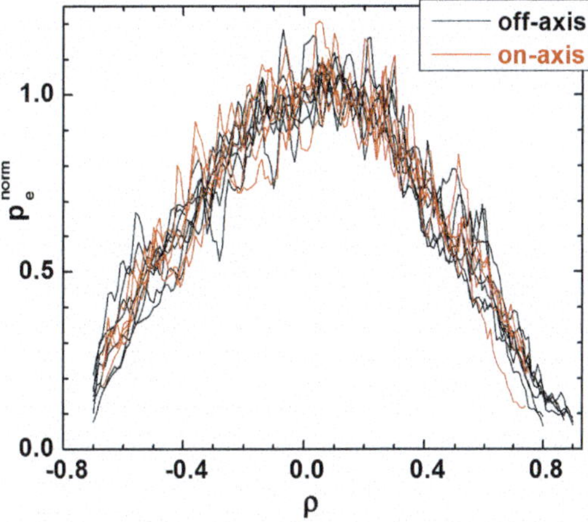

Fig. 1.4 The normalized electron pressure profiles $p_e = nT_e$ for both types of ECR heating in the stellarator TJ-II [25]

canonical profiles are used to determine the critical gradients in particle, heat and toroidal rotation fluxes. Canonical profiles of the electron temperature are found using Ohm's law. It is also assumed that the canonical temperature profiles for ions and electrons are the same. If the temperature gradient is larger than a certain critical threshold value, the corresponding term in the heat flux equation becomes dominant. If the gradient is less than the critical threshold value, this term in the heat flux equation is zero. This follows from the second law of thermodynamics, that heat cannot be transported against the temperature gradient. To describe the particle flux, we use the canonical pressure profile. This automatically leads to the description of thermal convection effects. In such a form the transport model is suitable to describe Ohmic and L-mode plasmas.

To describe shots with improved confinement of energy and particles, we introduce the concept of the second critical gradient. In Ohmic and L-mode plasmas the temperature gradient is generally larger than the critical threshold, but less than the second critical gradient threshold. If the temperature gradient in a certain region of the plasma is greater than the second critical gradient threshold, the plasma in this region "forgets" the canonical profile, and the corresponding term in the flux of heat or particles vanishes. This means that a transport barrier is formed. In the rest of the plasma cross section the plasma continues to "remember" the canonical profile, and the structure of heat and particle fluxes is preserved. In order to describe the effect of forgetting, the multipliers, which depend non-linearly on the gradients, are introduced into the heat and particle fluxes. If the gradients of the plasma parameters exceed the threshold of second critical gradient, a bifurcation takes place in the equations of heat and/or particle diffusion. New solutions correspond to plasma with an external or internal transport barrier. The second critical gradient threshold for the external transport barrier can be considered as well established, but work to determinate the second critical gradient for internal transport barriers still ongoing. Apparently, the internal transport barriers (Internal Transport Barrier—ITB) initially appear in the vicinity of resonant surfaces $q = m/n$ with small values of

m and n: $m/n = 3/2,\ 2/1,\ 5/2,\ 3/1$. The width of the ITB usually increases with additional plasma heating. This may be a consequence of changes in the current profile inside and around the transport barrier; however, the details of this effect have not been studied.

Note that the formation of a transport barrier presents similarities to the non-linear Hooke's law. When a solid body is stretched, the elastic force is proportional to the extension until the limit of plasticity is reached. If this limit is exceeded, the force decreases, and the solid body "forgets" its initial state. In plasma, this process corresponds to the formation of a transport barrier.

A phenomenological transport model determines the ambiguity in the choice of a model for the flux of heat and particles. This choice is justified *aposteriori* after comparing the calculated results with experimental data. However, several additional conditions reduce the possibility of large error in the choice of model. Since variations in the shape of the canonical profile in the L-mode has only a small effect on the results of the calculations, due to the fact that in the L-mode the temperature profile is markedly different from the canonical profile.

Furthermore, the selected transport model is "stiff", as the transport coefficient (stiffness), located before the difference of temperature gradient and critical gradient threshold is relatively large. In constructing our model, we adhere to the "simplicity principle." This means that the model should contain a minimum 'fudge' factors. Furthermore, a conventional L-mode model should be linear relative gradients of required functions such as temperature and plasma density. These limitations lead to an almost unequivocal choice in the structure of the expressions for heat and particles fluxes. Details related to the transport model are discussed in Chap. 5 and 6.

References

1. Coppi, B.: Nonclassical transport and the 'principle of profile consistency'. Comment. Plasma Phys. Control Fusion. **5**, 261 (1980)
2. Esiptchuk, Yu.V., Razumova, K.A.: Investigation of plasma confinement on Soviet tokamaks. Plasma Phys.Control Fusion. **28**, 1253 (1986)
3. Dnestrovskij, Yu.N., Kostomarov, D.P.: Numerical Simulation of Plasmas. Springer, Berlin (1986)
4. Dnestrovskij, Yu.N., Razumoval, K.A.: Self-consistency of pressure profiles in tokamaks. Nucl. Fusion. 46, 953 (2006)
5. Taylor, J.B.: Relaxation of toroidal plasma and generation of reverse magnetic fields. Phys. Rev. Lett. **33**, 1139 (1974)
6. Kadomtsev, B.B.: Self-organization of tokamak plasma. Sov. J. Plasma Phys. **13**, 443 (1987)
7. Biskamp, D.: Natural current profiles in tokamaks. Comment. Plasma Phys. Control Fusion. **10**, 165 (1986)
8. Hsu, J.Y., Chu, M.S.: The tokamak equilibrium profile. Phys. Fluid. **30**, 1221 (1987)
9. Minardi, T., Weisen, H.: Stationary magnetic entropy in ohmic tokamak plasmas: experimental evidence from the TCV device. Nucl. Fusion. **41**, 113 (2001)
10. Minardi, E., Lazzaro, E.: Profile consistency based on the magnetic entropy concept: theory and observation. Nucl. Fusion. **43**, 369 (2003)
11. Hameiri, E., Bhattacharjee, A.: Entropy production and plasma relaxation. Phys. Rev. A. **35**, 768 (1987)

12. Phillips, L.: States of minimum dissipation in magnetohydrodynamics: a review. J. Plasma Phys. **56**, 531 (1996)
13. Zhang, C.: Relaxed states for Ohmically driven tokamaks with an arbitrary aspect ratio. Phys. Plasma. **11**, 1445 (2004)
14. Bhattacharyya, R., Janaki, M.S.: Dissipative relaxed states in two-fluid plasma with external drive. Phys. Plasma. **11**, 5615 (2004)
15. Kotschenreuther, M.: Quantitative predictions of tokamak energy confinement from first-principles simulations with kinetic effects. Phys. Plasma. **2**, 2381 (1995)
16. Nordman, H., Weiland, J., Jarmen, A.: Simulation of toroidal drift mode turbulence driven by temperature gradients and electron trapping. Nucl. Fusion. **30**, 983 (1990)
17. Waltz, R.E.: A gyro-Landau-fluid transport model. Phys. Plasma. **4**, 2482 (1997)
18. Idomura, Y.: Study of ion turbulent transport and profile formations using global gyrokinetic full-f Vlasov simulation. Nucl. Fusion. **49**, 065029 (2009)
19. Diamond, P.H.: Physics of non-diffusive turbulent transport of momentum and the origins of spontaneous rotation in tokamaks. Nucl. Fusion. **49**, 045002 (2009)
20. Aydemir, A.Y.: An intrinsic source of radial electric field and edge flows in tokamaks. Nucl. Fusion. **49**, 065001 (2009)
21. Callen, J.D.: Toroidal rotation in tokamak plasmas. Nucl. Fusion. **49**, 085021 (2009)
22. Honda, M., Takizuka, T., Fukuyama, A., Yoshida, M., Ozeki, T.: Self-consistent simulation of torque generation by radial current due to fast particles. Nucl. Fusion. **49**, 035009 (2009)
23. Wagner, F.: W7-AS: one step of the Wendelstein stellarator line. Phys. Plasma. **12** 072509 (2005)
24. Dnestrovskij, Yu.N., Melnikov, A.V., Pustovitov, V.D.: Approach to canonical pressure profiles in stellarators. Plasma Phys. Control Fusion. **51**, 015010 (2009)
25. Melnikov, A.V. et al.: Pressure profile shape constancy in L-mode stellarator plasmas. 34-th EPS Conference on Plasma Physics, Warshaw, ECA, vol 31F, Rep. P-2.060 (2007)
26. Dnestrovskij, Yu.N., Pereverzev, G.V.: Energy confinement in the T-10 tokamak and canonic profile models. Plasma Phys. Control Fusion. **30**, 1417 (1988)
27. Dnestrovskij, Yu.N. et al.: Transport model of canonical profiles for electron and ion temperatures in tokamaks. Nucl. Fusion. **31**, 1877 (1991)
28. Dnestrovskij, Yu.N., Lysenko, S.E., Tarasyan, K.N.: Improved confinement regimes within the transport model of canonical profiles. Nucl. Fusion. **35**, 1047 (1995)
29. Dnestrovskij, Yu.N., Dnestrovskij, A.Yu., Lysenko, S.E.: Self-organization of plasma in a tokamak. Plasma Phys. Rep. **31**, 529 (2005)
30. Dnestrovskij, Yu.N. et al.: Canonical profiles and transport model for the toroidal rotation in tokamaks. Plasma Phys. Control Fusion. **53**, 085025 (2011)

Chapter 2
Variational Principles for Canonical Profiles in a Tokamak

Abstract This Chapter is devoted to the variational formulation for the "canonical profiles" of the plasma temperature and pressure. The basis for the variational description is the functional for the magnetic energy associated with the plasma current together with the conditions for the conservation of the total plasma current and total magnetic flux. The variation of this functional leads to the Euler equation that defines the canonical profile. We start with a cylindrical plasma with circular cross section and then generalize to a toroidal plasma with arbitrary cross section. We also derive the variational formulation for the canonical profile of the toroidal rotation.

2.1 The Principle of Total Energy Minimum by Hsu and Chu

According to Hsu and Chu [1], let's introduce the polar coordinates system r, φ, z with z-axis which coincides with the main axis of the torus. The general form for magnetic field in a tokamak (axially symmetric torus) is as follows:

$$r\mathbf{B} = [\nabla\psi][\nabla\varphi][\nabla\varphi] \tag{2.1}$$

where ψ is a potential of poloidal magnetic field, $F = F(\psi)$ is a diamagnetic function ($F = rB_\varphi$). Grad-Shafranov two-dimensional equilibrium equation

$$\Delta^*\psi = -rj_\varphi = -(FF' + r^2 p') \tag{2.2}$$

determines the potential ψ distribution in space. Here $j_\varphi = j_\varphi(\psi, r)$ is a toroidal current density, $p = p(\psi)$ is a plasma pressure, $p' = dp/d\psi$,

$$\Delta^*\psi \equiv r\frac{\partial}{\partial r}\left(\frac{1}{r}\frac{\partial\psi}{\partial r}\right) + \frac{\partial^2\psi}{\partial z^2}. \tag{2.3}$$

The equation $\psi = \text{const}$ defines the magnetic surfaces.

It is assumed in this model that the relaxed equilibrium state of the plasma is determined by the minimum of the total plasma energy W (including magnetic and thermal energy)

Yu.N. Dnestrovskij, *Self-Organization of Hot Plasmas*, DOI 10.1007/978-3-319-06802-2_2,

$$W = \int_V dV \left\{ \left[F^2 + (\nabla \psi)^2 \right] / (2r^2) + \frac{3}{2} p \right\}, \tag{2.4}$$

while maintaining the total toroidal current

$$I = \frac{1}{2\pi} \int_V \frac{(F F' + r^2 p')}{r^2} dV \tag{2.5}$$

and the equilibrium condition (2.2). The independent variable in the variation problem is the variable ψ.

We will solve the variation problem (2.2, 2.4, 2.5) with Lagrange method. Let's introduce the extended functional

$$W_{ex} = W - 2\pi\lambda I \tag{2.6}$$

and consider the problem of the unconditional extremum of the functional (2.6). The first variation of (2.6) must be equal to zero in the extremum point:

$$\delta W_{ex} = \delta W - 2\pi\lambda\delta I = \int_V dV \, \delta\psi \left[\frac{FF' - \Delta^* \psi}{r^2} + \frac{3}{2} p' - \lambda \left(\frac{F'F' + FF''}{r^2} + p'' \right) \right] = 0. \tag{2.7}$$

As ψ is the independent variable and $\delta\psi$ is an arbitrary increment, so we obtain a two-dimensional Euler equation

$$\frac{FF' - \Delta^* \psi}{r^2} + \frac{3}{2} p' - \lambda \left(\frac{F'F' + F F''}{r^2} + p'' \right) = 0 \tag{2.8}$$

Let's substitute the equilibrium Eq. (2.2) to Eq. (2.8); then we reduce the Euler equation to the form

$$\left[2FF' - \lambda (F F')' \right] / r^2 + \left[\frac{5}{2} p' - \lambda p'' \right] = 0. \tag{2.9}$$

The second term in square brackets in (2.9) is permanent on the magnetic surface. The first term is permanent only if it is equal to zero. As a result, two-dimensional Eq. (2.9) is divided into two independent one-dimensional equations

$$2FF' - \lambda (F F')' = 0, \quad \frac{5}{2} p' - \lambda p'' = 0. \tag{2.10}$$

For future convenience we denote $\frac{5}{2\lambda}$ through λ. Then the solutions of Eq. (2.10) take the form

$$FF' = C_F \exp\left(\frac{4}{5} \lambda \psi \right), \quad p' = C_p \exp(\lambda \psi). \tag{2.11}$$

After substitution (2.11) into (2.2), we obtain the canonical equilibrium equation

$$\Delta^* \psi = -rj_\varphi = C_F \exp\left(\frac{4}{5}\lambda\psi\right) - C_p r^2 \exp(\lambda\psi). \tag{2.12}$$

The boundary conditions should be formulated to determine the constants C_F, C_p and λ. For example the potentials on the magnetic axis O_1 and on the plasma boundary S and the total plasma current can be set: $\psi(O_1) = 0$, $\psi(S) = \psi_a$, $I = I_p$. Sometimes it is more convenient to set the values $I = I_p$, $\beta_p = \beta_p^0$, $q(0) = q_0$. Here β_p is the ratio of plasma pressure to the poloidal magnetic field pressure, $q = \delta\Phi/\delta\psi$ is the plasma safety factor, Φ is the toroidal magnetic flux inside the magnetic surface.

Later on we will need a canonical pressure profile $p_c = p_c(\psi)$. This profile, which determines the pressure profile in the relaxed state, should weakly depend on the boundary conditions. To do this, we require that $p_c(\psi)$ as $p_c'(\psi)$ in (2.11) exponentially dependent on ψ

$$p_c(\psi) = p_0 \exp(\lambda\psi). \tag{2.13}$$

Here, p_0 is the pressure on the magnetic axis, where $\psi = 0$.

If $\beta_p \sim 1$, the first term on the right side (2.2) is small compared to the second one and it can be omitted. After averaging the remaining part over the magnetic surface, we obtain

$$\langle j_\varphi \rangle = \langle r \rangle p_c' \propto \langle r \rangle p_c. \tag{2.14}$$

Here, the angle brackets denote the averaging over the magnetic surface S

$$\langle f \rangle = \int_S f\, dS \Big/ \int_S dS,$$

Taking the left hand side of (2.14) as a definition of the canonical current profile j_c, we obtain the connection between the canonical profiles of current and pressure

$$j_c \propto \langle r \rangle p_c. \tag{2.15}$$

Since $\langle r \rangle \approx R_0 + \Delta_S(\psi)$, where R_0 is a major radius of the torus, $\Delta_S(\psi)$ is a Shafranov shift and $\Delta_S(\psi) \ll R_0$, then $\langle r \rangle$ is little different from constant R_0, and the profiles j_c and p_c are close to each other.

In the approximation of a straight cylinder with a circular cross-section $\langle r \rangle = \text{const}$ and canonical profiles of current and pressure are the same: $j_c \propto p_c$. In this approximation, the canonical equilibrium equation (2.12) at $\beta_p \sim 1$ is as follows

$$\Delta\psi \equiv \frac{1}{\rho}\frac{d}{d\rho}\left(\rho\frac{d\psi}{d\rho}\right) = -C_p \exp(\lambda\psi), \tag{2.16}$$

where ρ is the radial coordinate in the cylinder. Its solution is the function

$$\psi = -\ln(1 + A_1\rho^2)^2, \tag{2.17}$$

so, the canonical profiles of current density and pressure are as follows

$$\frac{j_c(\rho)}{j_c(0)} = \frac{p_c(\rho)}{p_c(0)} = \frac{1}{(1 + A_2\rho^2)^2}, \tag{2.18}$$

where A_1 and A_2 are the constants determined by boundary conditions. It is convenient to define the boundary conditions for the function $\mu = 1/q = R_0B_\theta/\rho B_0$, where B_θ and B_0 are the poloidal and toroidal magnetic fields. If $\mu_0 = \mu(0)$ and $\mu_a = \mu(a)$, then $A_2 = \mu_0/\mu_a - 1$. The function $\mu_c(\rho)$ itself has the form

$$\mu_c(\rho) = \mu_0(1 + A_2\rho^2)^{-1}. \tag{2.19}$$

The comparison of (2.18) and (2.19) shows that

$$\frac{j_c(\rho)}{j_c(0)} = \left(\frac{\mu_c(\rho)}{\mu_0}\right)^2, \tag{2.20}$$

Thus, the variation problem with the functional (2.4) allows us to construct the canonical pressure profile (2.13), which coincides at $\beta_p \sim 1$ with canonical current profile.

Difficulties arise during the construction of transport model. Experiment shows that the characteristic relaxation times of temperature and current are very different. In a tokamak, the characteristic time of current profile relaxation may be ten times higher than the plasma energy confinement time. During the transition process of the plasma current evolution the current profile could be far from the canonical current profile and the equilibrium equation cannot be written in the form (2.12). As a result one has to abandon the use of a simple variational principle (2.4−2.5) and look for other variation approach for the design of transport models. This does not exclude the possibility that some of the conclusions of this section will be used further in this book.

2.2 The Principle of Minimum of the Plasma Current Magnetic Energy for a Circular Plasma Cylinder (Kadomtsev)

2.2.1 The Natural but Contradictive Statement of the Variational Problem

As before, we assume that the canonical profiles are the goals to the relaxation of plasma parameters. The toroidal magnetic field in a tokamak stabilizes the large-scale MHD instabilities, which characteristic times are much smaller than the relaxation times. Plasma, in turn, has little effect on the toroidal field, so this field can be removed from the energy reservoir defining the transport in plasma. In the previous section we see that the canonical pressure profiles associate with the canonical profiles of current by the equilibrium equation. Therefore, the thermal energy can also be excluded from the consideration of the problem of canonical profiles. As a result, we come to the following variation principle for a circular cylindrical plasma [2]: the relaxed plasma state is defined by the minimum of the magnetic energy of the toroidal current:

$$W_m = 2\pi \int_0^a \frac{B_\theta^2}{8\pi} \rho d\rho \tag{2.21}$$

provided that the current magnitude

$$I = 2\pi \int_0^a j\rho d\rho \tag{2.22}$$

and the magnetic flux

$$\Psi = 2\pi \int_0^a B_\theta d\rho \tag{2.23}$$

are conserved. Here ρ and θ are the radial and poloidal coordinates, a is a radius of plasma cylinder, $B_\theta = B_\theta(\rho)$ is the poloidal magnetic field, $j = j(\rho)$ is a current density.

For convenience let's introduce the dimensionless quantity

$$\mu = 1/q = R_0 B_\theta / (\rho B_0) \tag{2.24}$$

and accept it as an independent variable. Here B_0 is the magnitude of the toroidal magnetic field, R_0 is the equivalent of a major radius of the torus $(R_0 \gg a)$. The variation problem (2.21–2.23) will be solved by Lagrange method, for which we introduce the extended functional

$$W_{m,ex} = W_m + \lambda I + C\Psi. \tag{2.25}$$

For this functional the variation problem is reduced to the problem of the unconditional minimum. The additional assumption is introduced in [2] that in the vicinity of the extremum of the functional (2.25) the current density depends on μ only:

$$j = j(\mu). \tag{2.26}$$

What really lies behind this assumption will be discussed in the next Sect. 2.2.2.

Now we can find the first variation of the functional (2.25) and set it equals to zero:

$$\delta W_{m,ex} = 2\pi \int_0^a \rho d\rho \left(\frac{B_0^2}{4\pi R_0^2} \rho^2 \mu + \lambda \frac{dj}{d\mu} + \frac{CB_0}{R_0} \right) \delta\mu = 0. \qquad (2.27)$$

Hence we obtain the Euler equation

$$\frac{B_0^2}{4\pi R_0^2} \rho^2 \mu + \lambda \frac{dj}{d\mu} + \frac{CB_0}{R_0} = 0. \qquad (2.28)$$

We assume that the desired solution $\mu(\rho)$ of equation (2.28) is monotonic along the radius, i.e. $\mu'(\rho) \neq 0$ at $\rho \neq 0$. Then

$$\frac{dj}{d\mu} = \frac{dj}{d\rho}\left(\frac{d\mu}{d\rho}\right)^{-1}. \qquad (2.29)$$

We also assume that at the extremal function (on the solutions of the Euler equation (2.28)) the Maxwell equation is satisfied. Its projection on the z-axis is given by

$$j = \frac{B_0}{\mu_{00} R_0} \frac{1}{\rho} \frac{d}{d\rho}(\rho^2 \mu) \qquad (2.30)$$

Then the Euler equation (2.28) is transformed as follows:

$$\rho^2 \frac{d\mu^2}{d\rho} + \lambda_1 \frac{d}{d\rho}\left[\frac{1}{\rho}\frac{d}{d\rho}(\rho^2 \mu)\right] + C_1 \frac{d\mu}{d\rho} = 0. \qquad (2.31)$$

The equation (2.31) is the equation of the second order with two, while non-defined, Lagrange parameters λ_1 and C_1. It can be reduced to the third order equation with one uncertain parameter

$$\frac{d}{d\rho}\left[\frac{\rho^2}{\mu'}\frac{d}{d\rho}\left(\mu^2 + \lambda_2 \frac{d\mu}{d\rho^2}\right)\right] = 0 \quad (\mu' \equiv d\mu/d\rho) \qquad (2.32)$$

In this form the equation can be easily integrated.

We need four boundary conditions to obtain a unique solution of (2.32). Symmetry condition at the magnetic axis leads to $\mu'(0) = 0$. The requirement of the current conservation means that $\mu(a) \equiv \mu_a = 0.2 I_p R_0/(a^2 B_0)$. The practical units that are used here: the plasma current I_p in MA, the length in m, the magnetic field B_0 in T. The rest two boundary conditions were chosen in [2] as follows: $\mu(0) = \mu_0 \sim 1$, $\mu(\rho) \to 0$ at $\rho \to \infty$. We will follow these conditions. We obtain the following set of four boundary conditions, collecting all the terms together:

$$\mu(0) = \mu_0 \sim 1, \quad \mu'(0) = 0, \quad \mu(a) \equiv \mu_a = 0.2 I_p R_0 / (a^2 B_0),$$

$$\mu(\rho) \to 0 \quad \text{at} \quad \rho \to \infty. \tag{2.33}$$

The solution of (2.32) satisfying the boundary conditions (2.33) will be called the canonical profile of the function μ and will be denoted as $\mu_c(\rho)$. After integration of (2.32), we obtain

$$\frac{\rho^2}{\mu'} \frac{d}{d\rho}\left(\mu^2 + \lambda_2 \frac{d\mu}{d\rho^2}\right) = C_2. \tag{2.34}$$

To determine the constant C_2, let us consider the behavior of the regular solutions of the equation (2.34) in the environment of the point $\rho = 0$. Under the first two conditions (2.33) in this environment

$$\mu = \mu_0 (1 + \alpha_2 \rho^2 + \alpha^4 \rho^4 + \cdots). \tag{2.35}$$

For solutions of the type (2.35), the left side of (2.34) tends to zero as $\rho \to 0$. Hence, $C_2 = 0$. As a result, the Euler equation is now can be written as

$$\frac{d}{d\rho}\left(\mu^2 + \lambda_2 \frac{d\mu}{d\rho^2}\right) = 0. \tag{2.36}$$

The second of the boundary conditions (2.33) holds for solutions of (2.36) automatically. Therefore, there are only three essential boundary conditions:

$$\mu(0) = \mu_0 \sim 1, \quad \mu(a) \equiv \mu_a = 0.2 I_p R_0 / (a^2 B_0),$$

$$\mu(\rho) \to 0 \quad \text{at} \quad \rho \to \infty. \tag{2.37}$$

After integration (2.36), we obtain:

$$\mu^2 + \lambda_2 \frac{d\mu}{d\rho^2} = C_3. \tag{2.38}$$

At $\rho \to \infty$ the left side of (2.38) tends to zero by virtue of the third boundary condition (2.37). Hence, $C_3 = 0$. As a result, the Euler equation (2.38) becomes:

$$\lambda_2 \frac{d\mu}{\mu^2} = -d\rho^2 \tag{2.39}$$

Its solution is the function

$$\mu_c(\rho) = \frac{\mu_0}{1 + \rho^2 / a_j^2} \tag{2.40}$$

where $a_j = a\,[\mu_a/(\mu_0 - \mu_a)]^{1/2}$ is the so called *plasma current radius*, $\lambda_2 = \mu_0 a_j^2$. The function (2.40) will be called *Kadomtsev canonical profile* and will be denoted as $\mu_c^K(\rho)$. Using (2.30), we find the Kadomtsev canonical profile for the current density

$$j_c^K = j_0 \left(\frac{\mu_c^K}{\mu_0} \right)^2 = \frac{j_0}{(1+\rho^2/a^2)^2}. \tag{2.41}$$

By (2.15), for a circular cylinder the canonical profiles of current and pressure are the same. Therefore

$$p_c^K = p_0 \left(\frac{j_c^K}{j_0} \right) = \frac{p_0}{(1+\rho^2/a^2)^2}. \tag{2.42}$$

In subsequent chapters, we will need a dimensionless relative gradient of $\mu_c(\rho)$

$$-R_0 \frac{d\mu_c^K/d\rho}{\mu_c^K} = 2\frac{R_0\rho}{a_j^2}\frac{1}{1+\rho^2/a_j^2}. \tag{2.43}$$

Since the solution of (2.36) – (2.37) is found, it is easy to reformulate the last of boundary conditions (2.37), replacing it with the boundary conditions on the surface of the plasma. It is convenient to introduce the surface impedance in the form $X = i_a / 2\mu_a$, where i is the dimensionless current

$$i = \frac{1}{\rho}\frac{d}{d\rho}(\rho^2 \mu) = 2\mu + \rho\mu', \quad i_a = i(a). \tag{2.44}$$

Using (2.40, 2.44), it is easy to find the impedance for the Kadomtsev canonical profile

$$X^K = \frac{\mu_a}{\mu_0}. \tag{2.45}$$

Thus, in the Kadomtsev problem for the equation (2.36) the equivalent boundary conditions can be used instead of the boundary conditions (2.37)

$$\mu(0) = \mu_0 \sim 1, \quad \mu(a) \equiv \mu_a = \frac{0.2 I_p R_0}{a^2 B_0}, \quad \frac{i_a}{2\mu_a} = \frac{\mu_a}{\mu_0}. \tag{2.46}$$

The last of the boundary conditions is a condition of the third kind, as it contains the unknown function μ and its derivative μ'. It is also unusual as it contains both the values of unknown function at the magnetic axis and at the plasma boundary. The boundary conditions (2.37) and the equivalent conditions (2.46) are naturally called "soft", as one of the boundary conditions (2.37) is stated at infinity, and it does not reflect the physical processes on the surface of the plasma.

2.2.2 The Adjusted Statement of the Variational Problem

In setting up the variation problem in the previous Sect. 2.2.1, we, following the work of [2], have been forced to assume (2.26) that $j=j(\mu)$. However, if j is a local current density, it must be associated with μ by a relation

$$j \sim \frac{1}{\rho}\frac{d}{d\rho}(\rho^2 \mu) = 2\mu + \rho\mu', \quad \mu' \equiv \frac{d\mu}{d\rho}. \tag{2.47}$$

We see that in this case, the current density j depends not only on μ, but also on the derivative μ'. Substituting (2.47) into the condition (2.22) and calculating the variation of the functional I, we see that it is equal to zero. Thus, this integral does not contribute to the Euler equation. This section is discussed how to remove the contradiction described.

Consider the following formulation of the variation problem for a circular plasma cylinder. Let $B_\theta = B_\theta(\rho)$ is the set of sufficiently smooth functions (admissible functions) vanished at $\rho=0$ (these are the functions describing the poloidal magnetic field). In parallel, we also introduce the dimensionless admissible functions $\mu=\mu(\rho)=R_0 B_\theta/(B_0\rho)$. Consider the problem of minimizing of the poloidal magnetic energy functional

$$F_1 = \frac{1}{8\pi}\int_0^a B_\theta^2 \rho d\rho \sim \frac{1}{8\pi}\frac{B_0}{R_0} W_m, \quad W_m = \int_0^a \mu^2 \rho^3 d\rho \tag{2.48}$$

with additional integral conditions [3]

$$J_1 = \int_0^a \mu^2 \rho d\rho = \text{const}, \tag{2.49}$$

$$J_2 = \int_0^a \mu \rho d\rho = \text{const} \tag{2.50}$$

and the boundary conditions

$$\mu(0) = \mu_0 \sim 1, \quad \mu(a) \equiv \mu_a = 0.2 I_p R_0/(a^2 B_0), \quad (0 < \mu_a < \mu_0). \tag{2.51}$$

Here I is a plasma current, and a is a plasma radius. To simplify the formulas below we omit the factor standing in F_1 (2.48) to W_m. Note that the integral that describes the plasma current is proportional to μ_a

$$I = \int_0^a j\rho d\rho \sim \mu_a. \tag{2.52}$$

and, in view of (2.46), is the same for all admissible functions. Therefore, the preservation of the total current is the result of the last of the boundary conditions (2.51). The meaning of (2.49) and (2.50) will be discussed below.

The problem (2.48–2.51) is equivalent to the problem of unconditional minimum of the extended functional

$$W_{m,ex} = \int_0^a \left(\mu^2\rho^2 + \lambda\mu^2 + C\mu\right)\rho d\rho, \tag{2.53}$$

where λ and C are Lagrange parameters. We find now the variation of the functional (2.53)

$$\delta W_{m,ex} = 2\int_0^a \delta\mu\left(\mu\rho^2 + \lambda\mu + \frac{C}{2}\right)\rho d\rho \tag{2.54}$$

and require that it vanish. Then we obtain the Euler equation

$$\mu\rho^2 + \lambda\mu + C/2 = 0. \tag{2.55}$$

As before, the solution of the Euler equation satisfying the boundary conditions (the canonical profile) will be denoted by the lower index "c". From (2.55) and (2.53), we obtain:

$$\mu_c = -\frac{C}{2(\rho^2 + \lambda)} = \frac{\mu_0}{(1 + \rho^2/a_j^2)}. \tag{2.56}$$

The boundary conditions (2.51) allow us to determine the parameters λ and C

$$\lambda = a_j^2 = \frac{a^2}{\mu_0/\mu_a - 1} > 0, \quad C = -\frac{2\mu_0}{\lambda}. \tag{2.57}$$

The solution (2.56) coincides with the solution (2.40) of the previous section, which we called Kadomtsev canonical profile. This solution has a remarkable property: the dimensionless canonical current (2.44)

$$i_c \equiv \frac{1}{\rho}\frac{d}{d\rho}(\rho^2\mu_c) = \frac{2\mu_c^2}{\mu_0} \tag{2.58}$$

is proportional to the square of the function μ_c. This feature justifies the choice of a first additional condition in the form of (2.49). Note that the current density (2.58) is not equal to zero at the plasma boundary, but has a pedestal.

The considered variation problem is strongly degenerated in the sense that neither the functional (2.48), nor the additional conditions (2.49) – (2.50) do not depend on the derivative $d\mu/d\rho = \mu'$. As a result, the Euler equation (2.55) is an algebraic rather than a differential equation of the second order, as it could be obtained in a non-degenerate case. The solution of the Euler equation contains only two undefined parameters λ and C, but we have four conditions: the boundary conditions (2.51) and conditions (2.49) and (2.50). In order the posed variation problem to be solvable, the conditions (2.49) and (2.50) have to be consisted with the boundary conditions (2.51). We obtain such consistency conditions if we substitute the found solution (2.56) to the conditions (2.49) – (2.50):

$$J_1 = \int_0^a \mu_c^2 \rho d\rho = \mu_0 \mu_a \frac{a^2}{2}, \tag{2.59}$$

$$J_2 = \int_0^a \mu_c \rho d\rho = \mu_0 \frac{a_j^2}{2} \ln\left(\frac{\mu_0}{\mu_a}\right). \tag{2.60}$$

Thus, the problem (2.48) – (2.51) is solvable, if the values of the integrals (2.49) – (2.50) are defined by (2.59) – (2.60).

The meaning of (2.50) can be understood, if we introduce the usual poloidal flux ψ: $B_\theta \propto \rho\mu = d\psi/d\rho$. The condition (2.50) requires the preservation of difference $\psi(a) - \psi(0)$, which is equivalent to the flux conservation for all admissible functions.

The second variation of the functional (2.53)

$$\delta^2 W_{m,ex} = 2\int_0^a (\delta\mu)^2 (\rho^2 + \lambda)\rho d\rho \tag{2.61}$$

is positive in view of (2.57). Thus, the solution (2.56) of the Euler equation (2.55) realizes the minimum of the functional (2.53).

The Euler equation (2.55) can be rewritten as

$$2\mu\rho^2 + \lambda \frac{d\mu^2}{d\mu} + C = 0 \tag{2.62}$$

or

$$2\mu\rho^2 + \lambda \frac{d\mu^2}{d\rho}\left(\frac{d\mu}{d\rho}\right)^{-1} + C = 0. \tag{2.63}$$

In order to be able to go from (2.62) to (2.63), it needs to add the monotony condition of the admissible functions $\mu(\rho)$ to the limitations (2.49–2.50). That means the condition $d\mu/d\rho \neq 0$ has to be satisfied in the region $(0 < \rho < a)$. It is evident that the function (2.56) satisfies this condition.

Let us substitute (2.56) and (2.58) into (2.63) and multiply it by $d\mu_c/d\rho$. Then we obtain the identity

$$\rho^2 \frac{d\mu_c^2}{d\rho} + \frac{\lambda\mu_0}{2} \frac{d}{d\rho}\left[\frac{1}{\rho}\frac{d}{d\rho}(\rho^2 \mu_c)\right] + C\frac{d\mu_c}{d\rho} = 0$$

We omit index "c" now, redefine the parameter λ and consider this expression as the equation concerning $\mu(\rho)$

$$\rho^2 \frac{d\mu^2}{d\rho} + \lambda \frac{d}{d\rho}\left[\frac{1}{\rho}\frac{d}{d\rho}(\rho^2\mu)\right] + C\frac{d\mu}{d\rho} = 0 \tag{2.64}$$

Now we will try to put the boundary conditions for the equation (2.64) to solve the problem with function (2.56). Equation (2.64) is a nonlinear second-order equation containing two parameters, so to obtain a unique solution four boundary conditions are required. Two of them should be taken from the boundary conditions (2.51), the third is a condition of the axial symmetry $\mu'(0) = 0$, and the fourth condition is the property (2.58) at the border. Thus, the total set of the necessary boundary conditions is as follows:

$$\mu_c(0) = \mu_0 > 0, \quad \mu'(0) = 0, \quad \mu_c(a) = \mu_a = 0.2I_p R_0/(a^2 B_0),$$

$$\frac{\mu_0 i_c(a)}{2\mu_c^2(a)} = 1 \tag{2.65}$$

The fourth of the boundary conditions, in view of (2.47) is the boundary condition of the third kind, containing a derivative of function μ at the edge of the plasma. The solution of (2.64) – (2.65) is the function (2.56). In the future, the equation (2.64) will be called the modified Euler equation. This equation has an important property: it does not contain any information about the integrand in (2.49), i.e. the function μ^2.

Let us make a few comments:

Remark 1 The Euler equation of the form (2.64) coincides with (2.31). This result justifies not quite consistent statement of variation problem in the previous section and the assumption (2.26). After comparison of Kadomtsev additional condition (2.22) with the condition (2.49), we can say that, instead of the current density j in (2.22) it is sufficient to use such a function of μ (in this case μ^2), which is close to j, but coincides with it only at the extremal function of the functional (2.25).This important result will be used below.

Remark 2 The additional condition (2.49) has the following remarkable property. On the one hand, it is really the restriction on the class of admissible functions and contributes to the Euler equation. On the other hand, according to (2.58), this condition for the extremal function μ_c has a form of the current conservation

$$J_1 = \int_0^a \mu_c^2 \rho d\rho = \frac{\mu_0}{2}\int_0^a i_c \rho d\rho = \text{const.} \tag{2.66}$$

However, such a condition is already used in the boundary conditions (2.51). It follows that the condition (2.49) is degenerated on the extremal, and so it is weak in the environment of the extremal. This property of the problem (2.48) – (2.51) makes it selected of other possible variation problems for the functional (2.48) with integral constraints. One can expect that it defines the lower boundary of the functional (2.48) with respect to other reasonable integral limitations that may be used instead of (2.49).

Remark 3 In this section, we consider the problem of minimizing the magnetic energy of the plasma. Thermal energy is not part of this task, so the results of Sect. 2.1 can be adopted for it (for the plasma pressure), and, in particular, the formula (2.18). For our purposes, it can be written as follows:

$$\frac{p_c(\rho)}{p_c(0)} = \frac{i_c(\rho)}{i_c(0)} = \frac{1}{\left(1+\rho^2/a_j^2\right)^2} = \left(\frac{\mu_c(\rho)}{\mu_c(0)}\right)^2. \tag{2.67}$$

Let us consider now the total energy functional

$$E = 2\pi\left[\frac{1}{8\pi}\int_0^a B_\theta^2 \rho d\rho + \frac{3}{2}\int_0^a p\rho d\rho\right] \tag{2.68}$$

If we state the variation problem of the type (2.48 – 2.50) for this functional then the Euler equation will contain two unknown functions $\mu(\rho)$ and $p(\rho)$. The closure of the formulation of the problem will have to make an additional assumption about the relationship of these functions. We assume that the admissible functions are connected by the relation using as a basis the expression (2.67)

$$p(\rho) \propto \mu^2(\rho). \tag{2.69}$$

Then the second term in (2.68) takes the form of the integral (2.49). As a result, the Euler equation (2.55) and formulas (2.56 – 2.58) are saved.

A minimum of the total energy functional E (2.68) is determined by its value on the canonical profiles

$$E(\mu_c, p_c) = \left(\frac{B_0}{2R_0}\right)^2\left[W_m(\mu_c) + \frac{3}{4}a^4\frac{\mu_a^3}{\mu_0}\beta_\theta\right] \tag{2.70}$$

Here

$$W_m(\mu_c) = \int_0^a \mu_c^2 \rho^3 d\rho = \lambda(\mu_0 J_2 - J_1) > 0 \tag{2.71}$$

is the value of the magnetic energy over the canonical profile,

$$\beta_\theta = \frac{8\pi p_0}{B_{\theta a}^2}, \tag{2.72}$$

$B_{\theta a} = aB_0\mu_a/R_0$, p_0 is the pressure in the plasma centre, J_1 and J_2 are determined by formulas (2.59)–(2.60). The first term in (2.70) describes the minimum of magnetic energy, the second term is the heat energy. The magnetic energy in absolute value is limited by a given total current. In the adopted statement of the problem there is no limit on the absolute value of the thermal energy, so a minimum of heat energy is determined by the free parameter β_θ. The possibility of such a formulation of the problem is determined by the openness of the system, and the reasonableness of the statement that is by the existence of self organization plasma effects. The concept of openness means that the system is continuously absorbing and throwing out the heat energy and particles. The bound energy remaining in the plasma is determined by the dissipative properties of the system (thermal conductivity, radiation), so it cannot be described by the variation principle. To determine the evolution of the system one has to build the transport models, which are discussed in the following chapters.

Remark 4 It was shown above that for circular cylindrical plasma the Euler equation with the necessary boundary conditions can be represented in five different forms. First, it is an algebraic equation (2.55), which contains two parameters, with the boundary conditions (2.51). Second, it is the first order differential equation with a single parameter (2.39) and with the same boundary conditions. Third, it is the second order differential equation with a single parameter (2.36) and three boundary conditions (2.46). Fourth, it is a second-order differential equation (2.64) with two parameters and with four boundary conditions (2.65). Fifth and finally, it is the third-order differential equation (2.32) with one parameter and four boundary conditions (2.33) or (2.65). All listed forms of problem setting are equivalent in the sense that they have the same unique solution in the form of the canonical profile (2.56).

Remark 5 In subsequent chapters, we will need a canonical profile for temperature. It can be obtained on the basis of the following reasons [2]. In the relaxed quasi-stationary state the current profile $j(\rho)$ and electron temperature profile $T_e(\rho)$ are close to the canonical profiles $j_c(\rho)$ and $T_c(\rho)$

$$j(\rho) \approx j_c(\rho), \quad T_e(\rho) \approx T_c(\rho). \tag{2.73}$$

At the same time the profiles $j(\rho)$ and $T_e(\rho)$ are related by Ohm's law

$$j(\rho) \propto T_e^{3/2}(\rho). \tag{2.74}$$

According to (2.67), (2.73) and (2.74), we have

$$p_c(\rho) = n_c(\rho)T_c(\rho) \propto j_c(\rho) \propto T_c^{3/2}(\rho). \tag{2.75}$$

It follows that

$$T_c \propto j_c^{2/3} \propto p_c^{2/3}, \quad n_c \propto T_c^{1/2} \propto j_c^{1/3}. \tag{2.76}$$

For the relative dimensionless gradients we obtain

$$\frac{R_0 T_c'}{T_c} = \frac{2}{3}\frac{R_0 p_c'}{p_c}, \quad \frac{R_0 n_c'}{n_c} = \frac{1}{3}\frac{R_0 p_c'}{p_c}, \quad T' = \frac{dT}{d\rho}. \tag{2.77}$$

Thus, the canonical profile of the electron temperature for a circular cylindrical plasma has the form

$$T_c(\rho) = \frac{T_c(0)}{(1+\rho^2/a_j^2)^{4/3}}. \tag{2.78}$$

We accept below that the canonical profiles for ions and electrons are the same.

2.3 Canonical Profiles for Toroidal Plasma with Arbitrary Cross-Section

Suppose that for a given distributions of current and pressure in a toroidal plasma with arbitrary cross section the problem of equilibrium with the given boundary conditions (the Grad-Shafranov equation for the function of poloidal magnetic flux ψ) is solved [4]. Then the equation $\psi = \text{const}$ determines the magnetic surfaces.

Denote by r, φ, z the polar coordinates with axis coinciding with the symmetry axis of torus. Let us introduce the natural coordinates ρ, θ, ζ, where ρ is the coordinate of magnetic surface defined by the toroidal magnetic field flux $\mathbf{\Phi}$ inside the magnetic surface with the cross section S.

$$\rho = \left(\frac{\Phi}{\pi B_0}\right)^{1/2}, \quad 0 < \rho < \rho_{max}, \quad \Phi = \int_S \mathbf{B}\cdot d\mathbf{S}, \quad \rho_{max} = \left(\frac{\Phi_{max}}{\pi B_0}\right)^{1/2} \tag{2.79}$$

Here θ is the poloidal angle, $\zeta = r\,\varphi$, the vector $\mathbf{dS}$ is perpendicular to the cross section S, $\mathbf{\Phi}_{max}$ is the total toroidal flux through the plasma cross section. Below B_0 is the vacuum toroidal field in the center of the chamber, $A = R_0/a$ is the aspect ratio, R_0 and a are the major and minor plasma radii. We will use the notation $h(\rho_{max}) = h_a$ for the boundary values of surface function $h(\rho)$. The angle brackets $<\cdots>$ denote the averaging operation over the magnetic surface

$$< f > = \frac{2\pi}{V'}\int_0^{2\pi} \sqrt{g}\, f\, d\theta, \tag{2.80}$$

where g is the determinant of the metric tensor of the accepted system of coordinates:

$$\sqrt{g} = r\frac{D(r,z)}{D(\rho,\theta)}, \quad V' \equiv \frac{\partial V}{\partial \rho} = 2\pi \int_0^{2\pi} \sqrt{g}\, d\theta, \tag{2.81}$$

$V(\rho)$ is the volume of the plasma inside the magnetic surface. We also introduce a surface function

$$\mu = \frac{\partial \psi}{\partial \Phi} = \frac{1}{2\pi B_0 \rho} \frac{\partial \psi}{\partial \rho}. \tag{2.82}$$

As it is known, in a toroidal plasma the local poloidal magnetic field B_θ^l and local current density j_φ^l are not the surface functions. In particular,

$$B_\theta^l = \frac{|\nabla \rho|}{2\pi r} \frac{\partial \psi}{\partial \rho}. \tag{2.83}$$

Instead of the local poloidal magnetic field B_θ^l and local current density j_φ^l we introduce the averaged poloidal field $B_\theta = \langle B_\theta^l \rangle$ and averaged current density $j = \langle j_\varphi^l \rangle$ using (2.80). The surface functions $\mu \propto B_\theta/\rho$ and j are connected by the Maxwell equations. It is convenient to introduce the dimensionless current density i, proportional to the averaged current density j and similar to (2.44). In the adopted coordinates the relationship between the dimensionless current density and function μ is as follows

$$i = \mathrm{rot}_\varphi(\rho\mu) = \frac{1}{V'} \frac{\partial}{\partial \rho}(V' G \rho \mu), \tag{2.84}$$

where V defined by (2.81) and

$$G = G(\rho) = R_0^2 \left\langle \frac{(\nabla \rho)^2}{r^2} \right\rangle \tag{2.85}$$

are the metric coefficients. Note that the coefficient G is dimensionless. In a circular cylindrical plasma $G \equiv 1$, $V' = (2\pi R) \cdot (2\pi \rho)$.

The expression for the magnetic energy of the poloidal magnetic field has the form:

$$W_m = 2\pi R_0 \frac{1}{8\pi} \int_{S_{max}} B_\theta^2 dS. \tag{2.86}$$

Here S_{max} is the total cross-section of the plasma. Below the multiplier $2\pi R_0$, which appears due to integration over the toroidal coordinate, is omitted. Using relations (2.82)−(2.83), (2.85), it is easy to lead the functional (2.86) to the one-dimensional integral

$$W_m = \frac{B_0^2}{2\pi R_0^2} \int_0^{\rho_{max}} V' G \mu^2 \rho^2 d\rho \tag{2.87}$$

We turn now to the formulation of the variation problem. Let $\mu(\rho)$ be a set of sufficiently smooth functions on the interval $(0<\rho<\rho_{max})$, satisfying the boundary conditions

$$\mu(0) = \mu_0 \sim 1, \quad \mu(a) = \mu_a, \quad (0 < \mu_a < \mu_0). \tag{2.88}$$

Here μ_a is defined by the equilibrium solution of the problem mentioned at the beginning of the section. As in the previous section, we assume that the canonical profile in a tokamak plasma is determined by the minimum of the functional of the poloidal field magnetic energy (2.87) with additional conditions

$$J_1 = \int_0^{\rho_{max}} V'i(\mu)d\rho = \text{const}, \tag{2.89}$$

$$J_2 = \int_0^{\rho_{max}} \mu\rho d\rho = \text{const}. \tag{2.90}$$

According to Remark 1 of Sect. 2.2.2, $i(\mu)$ is a smooth function of μ, satisfying on the extremal function the expression (2.84)

$$i(\mu_c) = i_c \equiv \frac{1}{V'}\frac{\partial}{\partial\rho}(V'G\rho\mu_c). \tag{2.91}$$

Condition (2.90) describes the conservation of the poloidal magnetic flux for admissible functions. In fact, substituting (2.82) into (2.90), we obtain

$$J_2 = \frac{1}{2\pi B_0}\int_0^{\rho_{max}} \frac{d\psi}{d\rho}d\rho = \frac{1}{2\pi B_0}\left[\psi(\rho_{max}) - \psi(0)\right] = \text{const}. \tag{2.92}$$

We apply Lagrange method to the problem of (2.87)–(2.90). For this we introduce the extended functional

$$W_{m,ex} = \frac{B_0^2}{2\pi R_0^2}\int_0^{\rho_{max}} V'\left[G\mu^2\rho^2 + \lambda i(\mu) + C\frac{\mu\rho}{V'}\right]d\rho \tag{2.93}$$

and consider the problem of the unconditional extremum for it. For this we find the first variation of the functional (2.93) with respect to variable μ and put it to zero

$$\delta W_{m,ex} = \frac{B_0^2}{2\pi R_0^2}\int_0^{\rho_{max}} V'\delta\mu\left[2G\mu\rho^2 + \lambda\frac{\partial i(\mu)}{\partial\mu} + C\frac{\rho}{V'}\right]d\rho = 0. \tag{2.94}$$

Hence we obtain the Euler equation

$$2G\mu\rho^2 + \lambda\frac{\partial i(\mu)}{\partial\mu} + C\frac{\rho}{V'} = 0. \quad (2.95)$$

We now require in addition that the admissible functions $\mu(\rho)$ are monotonic. It's enough that they satisfy a condition

$$\mu'(\rho) \neq 0 \text{ at } 0 < \rho < \rho_{max} \quad (2.96)$$

Then

$$\frac{\partial i(\mu)}{\partial\mu} = \left(\frac{\partial i(\mu)}{\partial\rho}\right)\left(\frac{\partial\mu}{\partial\rho}\right)^{-1}. \quad (2.97)$$

Substituting (2.97) into (2.95), using (2.91) and carrying out the transformations, we obtain instead of (2.95)

$$\rho^2 G\frac{\partial\mu^2}{\partial\rho} + \lambda\frac{\partial}{\partial\rho}\left[\frac{1}{V'}\frac{\partial}{\partial\rho}(V'G\rho\mu)\right] + C\frac{\rho}{V'}\frac{\partial\mu}{\partial\rho} = 0. \quad (2.98)$$

Equation (2.98) generalizes the modified Euler equation (2.64) in the case of the toroidal plasma with arbitrary cross section. This equation is a nonlinear second-order equation containing two parameters C and λ. To determine the unique solution four boundary conditions are required, which are similar to conditions (2.65), but taking into account the toroidal geometry:

$$\mu(0) = \mu_0 > 0, \quad \mu'(0) = 0, \quad \mu(\rho_{max}) = \mu_a, \quad \frac{\mu_0 i_a}{2}\Big/ G_a\mu_a^2 = U. \quad (2.99)$$

Here $U \sim 1$ is a constant. The value of U, equal to one in the case of a circular cylinder, is now to be determined by some additional considerations that help to highlight the Kadomtsev type solution, at least approximately satisfying the condition (2.58). To do this, let's consider the function

$$Z(\rho) = \frac{\mu_0 i_c(\rho)}{2}\Big/ G(\rho)\mu_c^2(\rho). \quad (2.100)$$

Here $\mu_c(\rho)$ is the solution of (2.98)−(2.99) for some U and i_c is determined by (2.91). On the boundary of plasma $Z(\rho_{max}) = U$ due to (2.99). In the central part of the plasma the aspect ratio of magnetic surfaces is large, their cross section is close typically to the circle and $G \approx 1$. Therefore, and due to (2.58), $Z(0) \approx 1$. Calculations show that for $U = 1$ everywhere in the interval $(0, \rho_{max})$ $Z(\rho) \geq 1$. Therefore, it has a maximum inside this interval. At the increase of U this maximum moves outside and at some value of $U=U_{opt}>1$ the function $Z(\rho)$ becomes monotonic over the whole interval. In this case, the maximum reaches the boundary $\rho=\rho_{max}$, and the derivative $\delta Z(\rho)/\delta\rho$ becomes zero at the boundary. In the cylindrical case $Z(\rho)\equiv 1$. So it is natural to suppose that in the toroidal case the function $Z(\rho)$ has to be monotonic and close to unity for Kadomtsev type solutions. Therefore, the value of U must be equal to U_{opt} and to define it, we can use the following condition:

$$\frac{\partial Z}{\partial \rho}(\rho=\rho_{max})=0. \tag{2.101}$$

The condition (2.101) should be added to the conditions (2.99) to determine the parameter U. The canonical current profile $i_c(\rho)=j_c(\rho)/j_c(0)$ is determined through $\mu_c(\rho)$ by the expression (2.84). The canonical pressure profile links with canonical current profile by (2.15). As $<r>$ is slowly changing function of ρ we replace it by constant and will determine the pressure canonical profile as follows

$$\frac{p_c(\rho)}{p_c(0)}=\frac{j_c(\rho)}{j_c(0)}\propto i_c(\rho)=\frac{1}{V'}\frac{\partial}{\partial\rho}(V'G\rho\mu_c). \tag{2.102}$$

In this approximation, the canonical profiles of pressure and current coincide. Canonical profiles of temperature and density of the plasma are determined by the formulas (2.76).

2.4 Examples

To illustrate the change of the canonical profiles with changing of plasma parameters we choose the typical discharges for three tokamaks: T-10, JET and MAST. T-10 has a circular cross section of the plasma and a large aspect ratio $A=R_0/a$. Typical discharge parameters are as follows:

$$R_0=1.5\,\text{m},\ a=0.3\,\text{m}\ (A=5), B_0=2.5\,\text{T}, I=0.25\ \text{MA}. \tag{2.103}$$

With these parameters, the cylindrical safety factor is equal to

$$q_{cyl}=\frac{5a^2B_0}{IR_0}=\frac{1}{\mu_{cyl}}=3.$$

The JET device (the largest in the world) has the elongated cross section and a moderate aspect ratio. Below we will use, as an example, the parameters of the discharge # 26087, which are as follows:

$$R_0=2.94\,\text{m},\ a=1.06\,\text{m}\ (A=2.8), B_0=3\text{T},$$

$$I = 3.2\ \text{MA},\ k = 1.65, \delta = -0.34, \tag{2.104}$$

where k and δ are the elongation and triangularity of plasma cross-section, $q_{cyl} = 1.67$. The value of q at the plasma boundary, obtained by the solution of the equilibrium equation by the three-moments method, equals to $q_a = 1/\mu_a = 3.9$.

The MAST device has an elongated cross-section and low aspect ratio. The parameters of a typical discharge # 11438 are

$$R_0 = 0.8\,\text{m},\ a = 0.6\,\text{m}\ (A = 1.35), B_0 = 0.47\,\text{T},$$

$$I = 0.62\ \text{MA},\ k = 1.8, \delta = 0.36,\ q_{cyl} = 1.2,\ q_a = 1/\mu_a = 13.1 \tag{2.105}$$

Let us look first at the behaviour of the metric coefficients. Figure 2.1 shows the normalized profiles of the coefficient V' for the discharges (2.103)–(2.105). For a circular cylindrical plasma $V' = \rho$ (dashed line in Fig. 2.1). It can be seen that by decreasing the aspect ratio the curves are increasingly deviate from the linear function. For tight aspect ratio tokamak MAST the function V' becomes non-monotonic, but of course it is always positive.

Figure 2.2 shows the behavior of the function G which characterizes the density of magnetic surfaces on the outer side of the torus. In a cylindrical plasma $G = 1$ for the whole cross section. With the decrease in the aspect ratio Shafranov shift increases, and the magnetic surfaces on the outer side of the torus are sealed. In MAST near the external plasma boundary the values of G are high: $G > 10$.

The boundary values $G(a)$ versus triangularity δ for MAST are shown in Fig. 2.3. At $\delta < 0$ the density of magnetic surfaces at the outer side of the torus diminishes and the values of $G(a)$ are reduced correspondingly.

We now turn to the canonical profiles. First, we compare the canonical profiles obtained in Kadomtsev approximation for circular plasma cylinder (2.40), and ones obtained by the solution of the problem (2.98)–(2.99) for the Euler equation. As an example we take the discharge parameters (2.103) of T-10. Figure 2.4 shows the canonical profiles $\mu_c^K(\rho)$ (2. 40) and $\mu_c(\rho)$ at $\mu_0 = 1$. It can be seen that they are close to each other. Figure 2.5 shows the profiles of the dimensionless relative pressure gradient

$$R_0 / L_{pc} = -\frac{R_0 p_c'}{p_c} \tag{2.106}$$

for canonical profiles in Kadomtsev approximation and in the general case. It is seen that the pressure gradients differ much more than the functions $\mu_c(\rho)$ shown in Fig. 2.4. Note that the values of ratios R_0/L_{pc} will play an important role in the transport models.

The following figures for the JET discharge (2.104) show the impact of the choice of the parameter U to the behaviour of the canonical profiles. Figs. 2.6 and 2.7 shows the canonical profiles μ_c and j_c for three values of the parameter U:

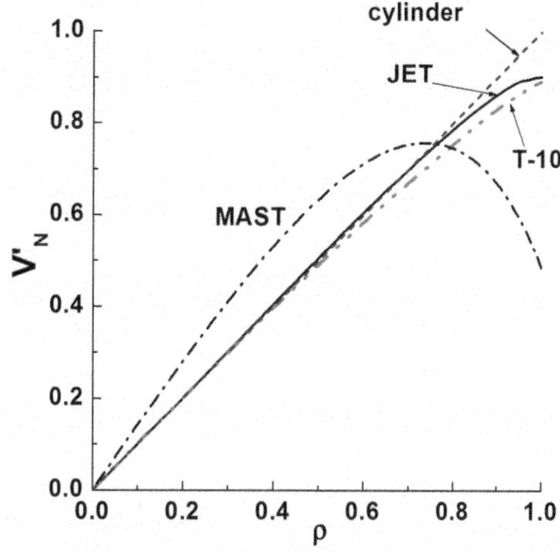

Fig. 2.1 The normalized profiles of the metric coefficient $V_N' = V' / (2\pi R_0 \cdot 2\pi\rho_{max})$ for discharges from T-10, JET and MAST devices with discharge parameters (1.103)–(1.105)

U=0.7, 1.3 (optimum) and 2. It is seen that the profiles of μ_c does not differ from each other radically at the change of U. They have the same boundary values and only derivatives are different on the boundary. The profiles of the canonical current j_c differ more because they have different boundary values. However, in all cases, they are monotonous. Figure 2.8 shows the profiles of the function $Z(\rho)$ for the same values of U. It can be seen that at U=0.7 the function $Z(\rho)$ has a maximum. In the second case (U=1.3) the condition (2.101) satisfies and in the third case (U=2) $Z(\rho)$ becomes monotonic and rapidly rises at the periphery. The profiles of critical gradients (2.106) for the same values of U are drawn in Fig. 2.9. The critical gradient for Kadomtsev canonical profile (2.42)

$$R_0 / L_{pc}^K = \frac{4R_0}{a_j^2} \frac{\rho}{1+\rho^2/a_j^2}. \tag{2.107}$$

is also shown here. It can be seen that at $U=U_{opt}$=1.3 the behaviour of R_0 / L_{pc}^K и R_0/L_{pc} is similar.

Let us turn to the canonical profiles of MAST. Figure 2.10 shows the profiles of the function $Z(\rho)$ for different values of U for the discharge # 11438 with parameters (2.105). In this case at U=1.5 the function $Z(\rho)$ has a maximum at ρ~0.8, but at $U=U_{opt} \approx 2$ the condition (2.101) satisfies. Corresponding profiles of critical gradients R_0/L_{pc} are shown in Fig. 2.11. Here, as in Fig. 2.9, the curves have a maximum at the edge with the values $U \sim U_{opt}$. The performed calculations show that the values of U_{opt} increase with decreasing of the aspect ratio and increasing of the cross section triangularity.

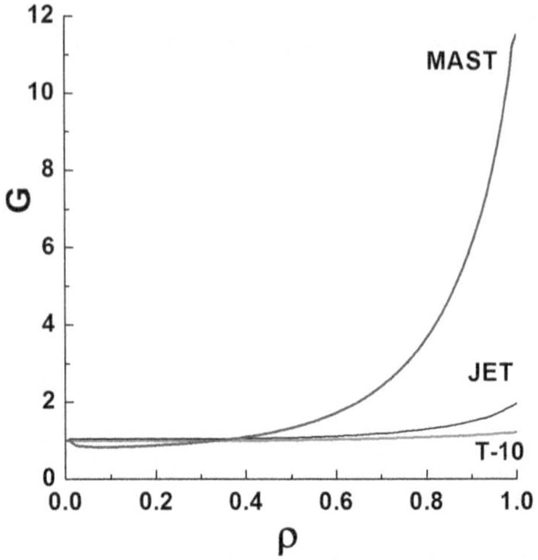

Fig. 2.2 The typical profiles of the metric coefficient G for T-10, JET and MAST devices with discharge parameters (1.103) − (1.105). This metric coefficient characterizes the density of the magnetic surfaces on the outer side of the torus

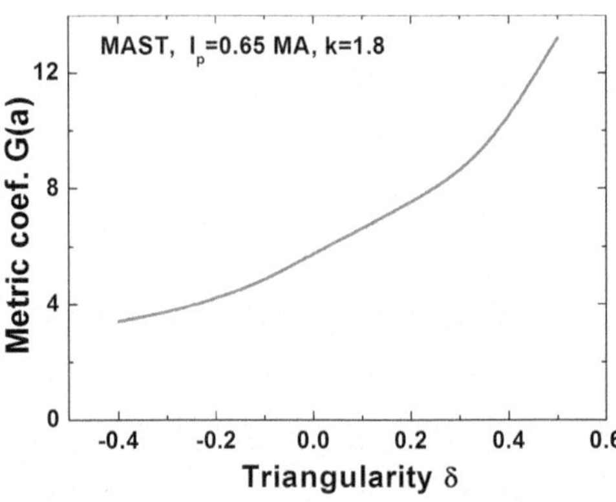

Fig. 2.3 The values of G at the MAST plasma boundary versus triangularity of plasma cross section at the elongation $k=1.8$. The density of magnetic surfaces on the outer side of the plasma increases at the increase of triangularity

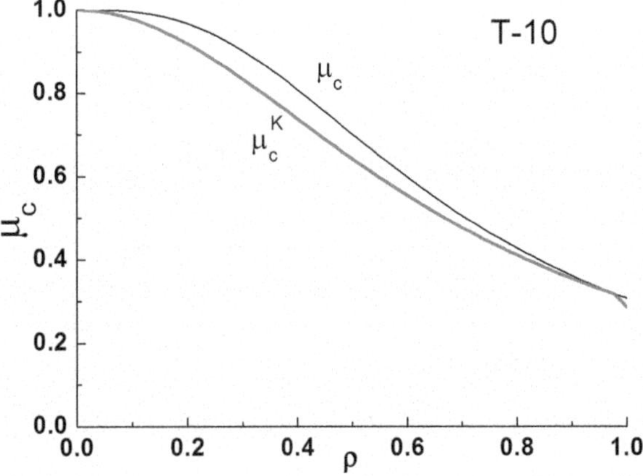

Fig. 2.4 The example of canonical profiles in general case and in Kadomtsev approximation for T-10 with plasma parameters (1.103)

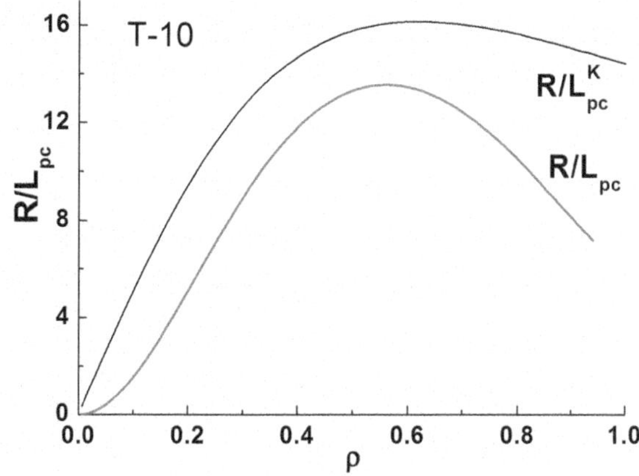

Fig. 2.5 The example of the dimensionless relative gradient profiles $R_0/L_{pc}=-R_0\, p_c'/p_c$ in general case and in Kadomtsev approximation for T-10 with plasma parameters (1.103)

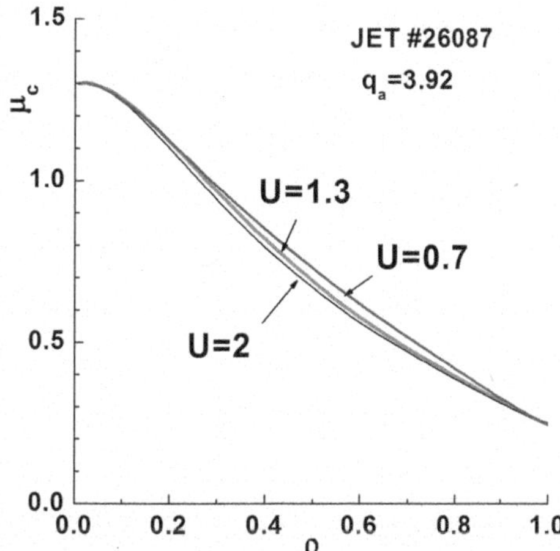

Fig. 2.6 The canonical profiles $\mu_c(r)$ for JET (discharge #26087) at different values of U

2.5 Canonical Profiles of the Toroidal Rotation

To determine the canonical profile of the plasma with toroidal rotation we use the procedure proposed in [1, 5] and discussed in Sect. 2.1. We first consider the equilibrium equation for plasma with rotation. For this we preset the toroidal rotation velocity υ_t as follows

$$\upsilon_t = \upsilon_t(\psi, r) = \omega r, \tag{2.108}$$

where $\omega=\omega(\psi)$ is the angular rotation frequency, r is a distance to the principal torus axis, ψ is the potential of the poloidal magnetic field, $\psi=$const at the magnetic surface. The equilibrium equation keeps the previous form (2.2)

$$\Delta^* \psi = -rj_\varphi = -(FF' + r^2 p'). \tag{2.109}$$

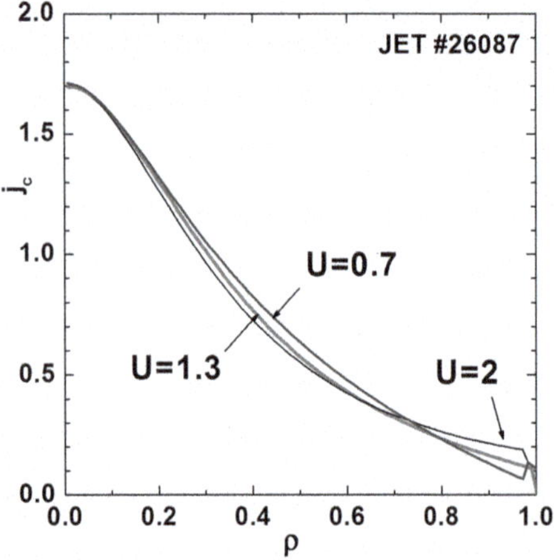

Fig. 2.7 The canonical profiles of current density $j_c(\rho)$ for JET at different values of U

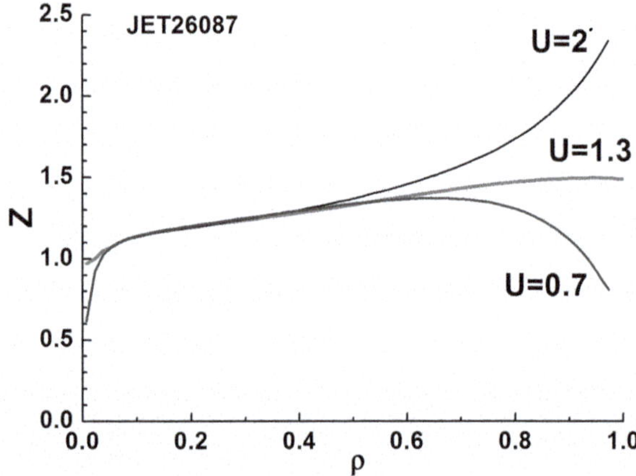

Fig. 2.8 Profiles of the function Z for JET at different values of U. It can be seen that the value of $U=1.3$ is optimal

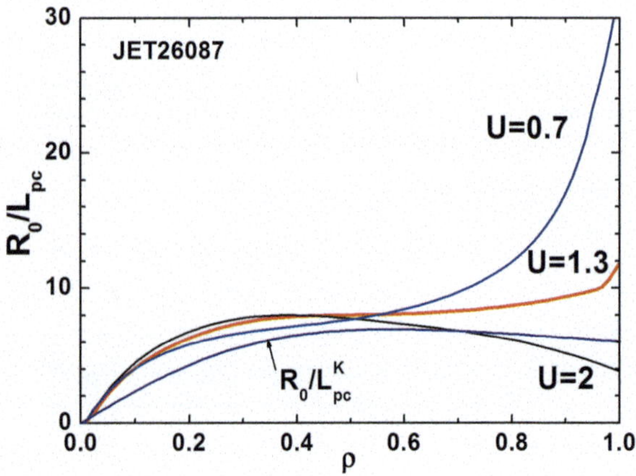

Fig. 2.9 Profiles of the relative pressure gradients R_0/L_{pc} for JET at different U. At the optimal value of $U=1.3$ the profile R_0/L_{pc} is almost flat at $\rho>0.3$. The profile R_0/L^K_{pc} in Kadomtsev approximation is also shown

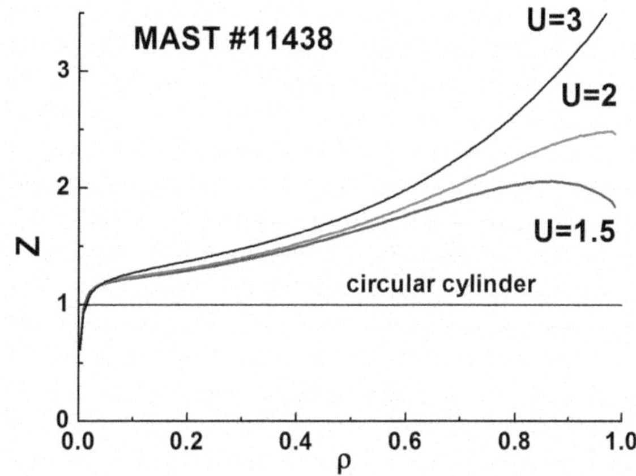

Fig. 2.10 Profiles of the function $Z(\rho)$ for MAST (discharge # 11438) at different values of U. It can be seen that the value of $U=2$ is close to the optimal one. In a circular cylindrical plasma $Z(\rho) \equiv 1$

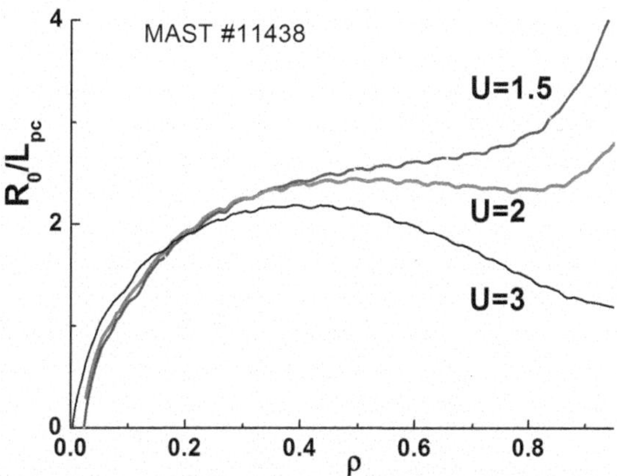

Fig. 2.11 Profiles of the relative pressure gradients R_0/L_{pc} for MAST (discharge # 11438) at different values of U

However, unlike the case of non rotated plasma, the plasma pressure p now depends on two variables and impermanent on the magnetic surface

$$p = p(\psi, r). \tag{2.110}$$

Let's recall that the bar (′) in (2.109) means the derivative with respect to ψ. The dependence of p on r must satisfy the condition (centrifugal force)

$$\frac{\partial p}{\partial r} = \frac{\rho_m v_t^2}{r} = \rho_m r \omega^2. \tag{2.111}$$

Here ρ_m is the hydrodynamic plasma density ($\rho_m = n\, m_i$, n is the plasma density, m_i is the ion mass). We assume that the kinetic energy of plasma rotation is much lower than the thermal energy

$$\frac{\rho_m \upsilon_t^2}{2} \ll p, \quad \text{or}\, \upsilon_t^2 \ll \upsilon_T^2, \quad M^2 = \frac{\upsilon_t^2}{\upsilon_T^2} \ll 1. \tag{2.112}$$

Here υ_T is the thermal velocity of the ions, M is Mach number. Now, as a function of $p(\psi, r)$, we can use a simple function that satisfies the condition (2.111)

$$p = p_0(\psi) + \frac{r^2}{2}\rho_m \omega^2 = p_0 + \frac{\rho_m \upsilon_t^2}{2}. \tag{2.113}$$

Let R_0 is a major radius of the plasma torus, then the function $(R_0^2/2)\,\rho_m\omega^2$ has the dimension of pressure. We denote it for the convenience through p_1. In this case

$$p = p_0 + \frac{r^2}{R_0^2} p_1, \quad p_1 = \frac{R_0^2}{2}\rho_m \omega^2. \tag{2.114}$$

By (2.112)

$$p_0 \gg p_1. \tag{2.115}$$

Therefore in the expression for p_1 (2.114) we can assume that $\rho_m = \rho_m(\psi)$. As a result, the equilibrium equation (2.109) takes the form

$$\Delta^* \psi = -r j_\varphi = -\left[FF' + r^2 \left(p_0 + \frac{r^2}{R_0^2} p_1 \right)' \right]. \tag{2.116}$$

We now turn to the variation problem. We define the canonical profiles of pressure, rotation, and function of the toroidal magnetic field F, as profiles that minimize the total energy of the plasma while maintaining the toroidal current and equilibrium conditions. The total energy is given by (compare with (2.4))

$$W = \int_V dV \left[\frac{F^2 + (\nabla \psi)^2}{2r^2} + \frac{3}{2} p + \frac{\rho_m \upsilon_t^2}{2} \right] \tag{2.117}$$

The total current is

$$I = \int_V \frac{1}{2\pi r} \frac{FF' + r^2 p'}{r} dV. \tag{2.118}$$

Lagrange extended functional is as follows

$$\Phi = W - 2\pi\lambda I. \tag{2.119}$$

Its first variation in the extremal point must be equal to zero

$$\delta\Phi = \delta W - 2\pi\lambda\,\delta I = \int_V dV\,\delta\psi\left\{\left[\frac{FF' - \Delta^*\psi}{r^2} + \frac{3}{2}\left(p_0 + \frac{r^2}{R_0^2}p_1\right)'\right] - \right.$$

$$\left. -\lambda\left[\frac{F'F' + FF''}{r^2} + \left(p_0 + \frac{r^2}{R_0^2}p_1\right)''\right]\right\} = 0 \tag{2.120}$$

Hence we obtain two-dimensional Euler equation

$$\left[\frac{FF' - \Delta^*\psi}{r^2} + \frac{3}{2}\left(p_0 + \frac{r^2}{R_0^2}p_1\right)'\right] - \lambda\left[\frac{F'F' + FF''}{r^2} + \left(p_0 + \frac{r^2}{R_0^2}p_1\right)''\right] = 0 \tag{2.121}$$

Using (2.116) it is converted to the form

$$\frac{2FF' - \lambda(FF')'}{r^2} + \left(\frac{5}{2}p_0' - \lambda p_0''\right) + \frac{r^2}{R_0^2}\left(\frac{5}{2}p_1' - \lambda p_1''\right) = 0. \tag{2.122}$$

The second term is permanent at a magnetic surface, but the remaining terms are variable with different dependencies on r. Hence we obtain three one-dimensional independent equations:

$$2FF' - \lambda(FF')' = 0,\quad \frac{5}{2}p_0' - \lambda p_0'' = 0,\quad \frac{5}{2}p_1' - \lambda p_1'' = 0. \tag{2.123}$$

Here, the first and second equations coincide with the corresponding equations for plasma without rotation (2.10). The third equation with respect to p_1 coincides with the equation for p_0. Thus, the canonical profile for the function $p_1 = \frac{R_0^2}{2}\rho_m\omega^2$ coincides with the canonical profile for the function p_0. Below, as in Sect. 1.1, we replace $5/(2\lambda)$ by λ. Then the solutions of equations (2.123) become as follows

$$FF' = C_F\exp(\tfrac{4}{5}\lambda\psi),\quad p_0' = C_{p0}\exp(\lambda\psi),\quad p_1' = C_{p1}\exp(\lambda\psi). \tag{2.124}$$

Substituting (2.124) into (2.116), we obtain the canonical equilibrium equation

$$\Delta^{*}\psi = -rj_{\varphi} = -\left[C_F \exp(\tfrac{4}{5}\lambda\psi) + C_p r^2 \exp(\lambda\psi)\left(p_0(0) + \tfrac{1}{2} r^2 \rho_{m0}\omega_0^2\right)\right]. \quad (2.125)$$

It is assumed that at the magnetic axis $\psi = 0$, $\rho_{m0} = \rho_m(0)$, $\omega_0 = \omega(0)$. The parameters C_F, C_p and λ are determined from additional conditions. For example

$$I = I_0, \quad \beta_p = \beta_p^0, \quad q(0) = q_0. \quad (2.126)$$

Integrating the last two Eq. (2.124), changing the notation of constants and leaving only the exponential parts of the solutions, we obtain, as in (2.13)

$$p_{c0} = C_p \exp(\lambda\psi), \quad p_{c1} = C_{p1} \exp(\lambda\psi). \quad (2.127)$$

Recalling that the canonical profiles are defined up to a multiplier, we obtain the following chain of equalities:

$$p_{c1} \propto p_{c0} \propto \rho_{mc}\omega_c^2 \propto n_c\omega_c^2 \propto n_c T_c. \quad (2.128)$$

Thus,

$$\omega_c^2 \propto T_c. \quad (2.129)$$

The canonical temperature profile is determined by (2.76), so

$$\omega_c \propto T_c^{1/2} \propto p_{c0}^{1/3}. \quad (2.130)$$

Later in the transport model the logarithmic derivatives of the canonical profiles will be used. By (2.130), they are linked by the following relations

$$\frac{\omega_c'}{\omega_c} = \frac{1}{3}\frac{p_{c0}'}{p_{c0}} = \frac{1}{2}\frac{T_c'}{T_c}. \quad (2.131)$$

Here the prime denotes the derivative with respect to the dimensional radial coordinate ρ, determining the magnetic surface ($0 < \rho < \rho_{max}$). The canonical current density and pressure profiles are calculated by the corresponding solutions of the problem (2.98, 2.99, 2.101) concerning $\mu = 1/q$ with one-dimensional Euler equation.

References

1. Hsu, J.Y., Chu, M.S.: The tokamak equilibrium profile. Phys. Fluids **30,** 1221 (1987).
2. Kadomtsev, B.B.: Self-organization of tokamak plasma. Sov. J. Plasma Phys. **13,** 443 (1987).
3. Dnestrovskij, Yu.N., Dnestrovskij, A.Yu. et al.: Variational Problems for the Canonical Profiles. Plasma Phys. Reports **34**(9), 794–797 (2008).
4. Dnestrovskij, Yu.N. et al.: Canonical Profiles in Tokamak Plasmas with an Arbitrary Cross Section. Plasma Phys. Reports **28**, 887 (2002).
5. Dnestrovskij, Yu.N. et al.: Canonical profiles and transport model for the toroidal rotation in tokamaks. Plasma Phys. Control Fusion **53**, 085025 (2011).

Chapter 3
A Possible Approach to the Canonical Profiles in Stellarators

Abstract The variational formulation for the canonical pressure profile in stellarators is discussed. The basis for this formulation is the functional of the total plasma energy (i.e. including the magnetic energy and thermal energy). The variation of this functional leads to one dimensional Euler equations for the pressure $p(\psi)$ and diamagnetic function $F(\psi)$ as a function of the potential ψ of the magnetic field. The solution of the equilibrium equation for ψ is found in the limit of low β. This then allows to determine the radial dependence of the canonical profile. The theoretical solution is compared with the experimental pressure profiles of ECR heated plasmas in the TJ-II stellarator.

The existence of canonical profiles in stellarators has still not been recognized by the scientific community as a whole and it was mentioned in the Introduction. However the example shown there displays that the effects of such a kind can exist in stellarators. In this chapter, we will take a look at a possible approach to the canonical pressure profiles in stellarators [1], using the variation principle proposed in [2] and discussed in Sect. 2.1.

3.1 Original Equations

We start with the energy integral

$$W = \int_V \left(\frac{\mathbf{B}^2}{2} + \frac{p}{\gamma - 1} \right) dV, \tag{3.1}$$

where $\mathbf{B}$ is the magnetic field, p is the plasma pressure, γ is the ratio of heat capacities and the integration is performed over the entire volume of the plasma. First, we integrate over the toroidal angle ζ, so W is transformed into an integral over the axially-symmetric torus inside the toroidal averaged plasma boundary.

In conventional stellarators with a flat circular axis all physical values can be represented as

Yu.N. Dnestrovskij, *Self-Organization of Hot Plasmas,* DOI 10.1007/978-3-319-06802-2_3,

$$f = \bar{f} + \tilde{f}, \tag{3.2}$$

where $\bar{f}(r, z)$ is the axially-symmetric (or toroidal averaged) part of f, and $\tilde{f}(r, \zeta, z)$ its oscillating part, (r, ζ, z) are usual cylindrical coordinates with the axis z directed along the main axis of the system. Similarly,

$$\mathbf{B} = \bar{\mathbf{B}} + \tilde{\mathbf{B}}, \tag{3.3}$$

where $\bar{\mathbf{B}}$ is the axially symmetric and $\tilde{\mathbf{B}}$ is the helical magnetic field

$$\bar{\mathbf{B}} = B_t \mathbf{e}_\zeta + \mathbf{B}_p, \tag{3.4}$$

$\mathbf{e}_\zeta$ is the unit vector in the toroidal direction, B_t is the toroidal component of $\bar{\mathbf{B}}$ and $\mathbf{B}_p$ is the poloidal magnetic field. It is shown in [3] that

$$\langle \mathbf{B}^2 \rangle = \langle \bar{\mathbf{B}}^2 - \tilde{\mathbf{B}}^2 \rangle, \tag{3.5}$$

where the brackets $\langle ... \rangle$ denote averaging over the volume between neighboring magnetic surfaces

$$\langle X \rangle \equiv \frac{d}{dV} \int X \, d^3 r. \tag{3.6}$$

Therefore

$$\int_V \mathbf{B}^2 dV = \int_{\bar{V}} \left(\bar{\mathbf{B}}^2 - \tilde{\mathbf{B}}^2 \right) d\bar{V}. \tag{3.7}$$

With this expression we get

$$W = \int_{\bar{V}} \left(\frac{\bar{\mathbf{B}}^2 - \tilde{\mathbf{B}}^2}{2} + \frac{\bar{p}}{\gamma - 1} \right) d\bar{V}, \tag{3.8}$$

where $\bar{p}$ is the axially symmetric part of p or the averaged pressure of the plasma. To simplify the notation we use p instead of $\bar{p}$ further.

According to (3.8) the integration is produced over a two-dimensional "quasi-tokamak" region $\bar{V}$. For stellarators the expression (3.8) is an approximation obtained by the stellarator expansion [3]. For tokamaks ($\tilde{\mathbf{B}} = 0$) this expression is exact.

3.2 Variation of Energy

If we want to calculate the variation of the energy storage, δW, we have to find $\delta \bar{\mathbf{B}}^2$. Axially symmetric poloidal field can be described in the form of $\mathbf{B}_p = \mathrm{rot} A_t \mathbf{e}_\zeta$. Therefore

$$\delta \frac{\mathbf{B}_p^2}{2} = \mathrm{div}\, \delta A_t \left(\mathbf{e}_\zeta \times \mathbf{B}_p \right) + j_\zeta \delta A_t, \tag{3.9}$$

where j_ζ is ζ—averaged toroidal component of the current density:

$$j_\zeta = \mathbf{e}_\zeta \cdot \mathrm{rot}\, \mathbf{B}_p. \tag{3.10}$$

Taking into account (3.9), we obtain from (3.8)

$$\delta W = \int_{\bar{V}} \left(B_t \delta B_t + j_\zeta \delta A_t + \frac{\delta p}{\gamma - 1} \right) d\bar{V}. \tag{3.11}$$

It is assumed that $\delta A_t = 0$ at the plasma boundary. The same assumption was used in [2]. We assume also that the helical field is fixed and does not vary it. Expression (3.11) also holds for tokamaks and stellarators. The difference for these systems is shown through various forms of A_t. For stellarators we have [3]:

$$\mathbf{B}_p = \frac{1}{2\pi} \left[\nabla \left(\psi - \psi_v \right) \times \nabla \zeta \right] \tag{3.12}$$

and

$$r A_t = \frac{\psi - \psi_v}{2\pi}. \tag{3.13}$$

Here $\psi(r, z) = \mathrm{const}$ describes the averaged magnetic surfaces and ψ_v is a poloidal flux of the helical field $\tilde{\mathbf{B}}$, which is equal to zero in tokamaks.

In Sect. 2.1 the energy is minimized, provided that the total value of the plasma current is invariant $\delta J = 0$. Following this approach, we consider the variation problem for the extended functional

$$\Phi = W - \lambda J, \tag{3.14}$$

where λ is the Lagrange multiplier. The current value is equal to

$$J = \int_S j dS = \int_{\bar{S}} j_\zeta d\bar{S} = \int_{\bar{V}} \frac{j_\zeta}{2\pi r} d\bar{V}, \tag{3.15}$$

where S is a cross-section of the plasma and $\bar{S}$ is its ζ—averaged projection, $d\bar{V} = 2\pi r\, d\bar{S}$. Then

$$\delta W - \lambda\delta J = \int_{\bar{V}} \left(B_t\, \delta B_t + \frac{j_\zeta}{r}\delta(rA_t) - \lambda\frac{\delta j_\zeta}{2\pi r} + \frac{\delta p}{\gamma - 1} \right) d\bar{V}, \tag{3.16}$$

and the variation principle together with (3.13) leads to the following Euler equation:

$$B_t\frac{\partial B_t}{\partial\psi} + \frac{j_\zeta}{2\pi r} - \frac{\lambda}{2\pi r}\frac{\partial j_\zeta}{\partial\psi} + \frac{p'(\psi)}{\gamma - 1} = 0. \tag{3.17}$$

Since we assume a helical field $\tilde{\mathbf{B}}$ fixed, the function ψ_V is not varied.

3.3 Application of Formulas to Stellarators

For stellarators with the condition $\left|\tilde{\mathbf{B}}/B_t\right| \ll 1$ we have [3]:

$$B_t = \frac{F(\psi)}{2\pi r} \tag{3.18}$$

and

$$j_\zeta = \frac{F F'(\psi)}{2\pi r} + 2\pi r\frac{dp}{d\psi}\left(1 + \Omega^0\right), \tag{3.19}$$

Here $\Omega^0 = \left\langle\tilde{\mathbf{B}}^2\right\rangle_\zeta / B_0^2$, where $\langle\ldots\rangle_\zeta$ denotes the toroidal averaging and B_0 is a toroidal magnetic field at the axis.

The value of Ω^0 can play an important role in special cases, when it has the order of reversed aspect ratio together with the strong asymmetry inside-outside on the magnetic surfaces. In the ordinary conditions $\Omega^0 \ll 1$ with little modulation on magnetic surfaces. Therefore, this value can be omitted in (3.19).

After substitution of (3.18) and (3.19) into (3.17), we obtain the Euler equation in the form

$$\frac{2FF' - \lambda(FF')'}{4\pi^2 r^2} + (p' - \lambda p'') + \frac{p'}{\gamma - 1} = 0, \tag{3.20}$$

where the prime denotes differentiation on ψ. Considering ψ and r^2 as independent variables, we get the final one-dimensional equations

$$2FF' - \lambda(FF')' = 0,\ \gamma p' - \lambda(\gamma - 1)\, p'' = 0 \tag{3.21}$$

that coincide with similar equations for a tokamak (2.10) at $\gamma = 5/3$, $\gamma/(\gamma-1) = 5/2$. It follows from (3.21) that for $\gamma = 5/3$

$$FF' = C_F \exp(2\psi/\lambda), \quad p' = C_p \exp(2.5\psi/\lambda). \tag{3.22}$$

With these functions the two-dimensional equilibrium equation for the stellarator [3]

$$\operatorname{div} \frac{\nabla(\psi - \psi_v)}{r^2} = -4\pi^2 \frac{dp}{d\psi} - \frac{FF'(\psi)}{r^2} \tag{3.23}$$

takes the form

$$\operatorname{div} \frac{\nabla(\psi - \psi_v)}{r^2} = -C_F \frac{\exp(2\psi/\lambda)}{r^2} - C_p \exp\left(\frac{2.5\psi}{\lambda}\right). \tag{3.24}$$

In this equation, the right-hand side, multiplied by r^2, is the same as the right-hand part of the equation for tokamaks (1.12) up to a value of $2.5/\lambda$ replacing by λ. It differs from the similar equations for the tokamak (1.12) with only a term ψ_v at the left side. The function ψ_v is determined by the helical field $\tilde{\mathbf{B}}$ and can be treated as known in (3.23) and (3.24).

Equation (3.24) is naturally called the canonical equilibrium equation for conventional stellarator. This is the equation of Poisson type and it requires the boundary conditions at the plasma boundary S, which is the magnetic surface

$$\psi(S) = \text{const} = \psi_b. \tag{3.25}$$

Equation (3.24) contains three parameters, C_F, C_p and λ, so we have to add three conditions for their determination. One of these conditions can be setting the value of ψ at the magnetic axis M_0

$$\psi(M_0) = \psi_0. \tag{3.26}$$

Instead, one can set the value of the rotational transform $\mu = \frac{d\psi}{d\Phi}$ at the magnetic axis. Here $\Phi = \int_{S_\psi} B_t dS$ is the toroidal magnetic flux on the magnetic surface ψ.

The second condition is to set the total current of the plasma:

$$J = \int_{S_\perp} \mathbf{j} \cdot d\mathbf{S}_\perp \tag{3.27}$$

In stellarators the total current is usually small, so we can assume that $J = 0$.

In experiments we usually know the value of the average β (of the ratio of plasma pressure to the magnetic field pressure), which is obtained from diamagnetic measurements. Therefore, as the third condition, the setting of the parameter

$$\beta_0 = 2p_0/B_0^2 \tag{3.28}$$

can be used, where p_0 and B_0 are the plasma pressure and toroidal field at the magnetic axis.

3.4 Canonical Pressure Profiles

The complete solution of two-dimensional canonical Eq. (3.24) will give us also the parameter λ, which appeared as the Lagrange multiplier in (3.14). Considering λ, we can find the profiles of F and p, which for a canonical equilibrium are determined by the one-dimensional Eq. (3.21). Replacing $2.5/\lambda$ by λ_1, we write (3.22) as follows:

$$p' = C_p \exp(\lambda_1 \psi). \tag{3.29}$$

As in Sect. 2.1, we shall refer to the exponential solution of (3.29) by the canonical pressure profile. It has the form

$$p_c = p_0 \exp(\lambda_1 \psi). \tag{3.30}$$

The solution of type (3.30) satisfies the boundary condition

$$\frac{1}{p}\bigg|_{\rho=\rho\max} \frac{dp}{d\psi}\bigg|_{\rho=\rho\max} = \lambda_1. \tag{3.31}$$

Suppose that on the magnetic axis at $\rho=0$, $\psi=\psi_0=0$. Then $p_0=p_c(\psi=0)$ is the pressure on the magnetic axis. Suppose also that the pressure at the plasma boundary is $p(\rho=\rho_{max})=p_a$. Then

$$p_0 \exp(\lambda_1 \psi_a) = p_a, \tag{3.32}$$

where $\psi_a=\psi(\rho=\rho_{max})$ is the value of ψ on the plasma boundary. From this we obtain the following expression for the parameter λ_1

$$\lambda_1 = -\frac{1}{\psi_a} \ln\left(\frac{p_0}{p_a}\right) \left(\frac{p_0}{p_a} > 1\right). \tag{3.33}$$

3.5 Approximate Solution of Equilibrium Equations

The general solution of Eq. (3.24) can be represented as

$$\psi = \psi_v + \psi_{ext} + \psi_{pl}, \tag{3.34}$$

where ψ_{ext} describes the contribution of the external poloidal field (for example the vertical quadrupole field) and ψ_{pl} is the contribution of the equilibrium currents in plasma. We assume that $\psi_{ext} = 0$, and consider the case when $|\psi_{pl}| \ll |\psi_v|$, which correspond to plasma with low β. Then

$$\psi \approx \psi_v. \tag{3.35}$$

This approach means that the shape of the magnetic surfaces is weakly perturbed by plasma.

Let's consider the simple case of a cylindrical plasma with a circular cross section of radius a. In this case $\psi_v = \psi_v(\rho)$, where ρ is the radial coordinate in plasma cross section. Then [3]

$$\frac{d\psi}{d\rho} = -2\pi\rho B_0\mu \tag{3.36}$$

where μ is a rotational transformation which, in general, may be approximated by the following expression

$$\mu = \mu_0 + (\mu_a - \mu_0)\xi^2, \tag{3.37}$$

where $\mu_0=\mu(0)$, $\mu_a=\mu(a)$, $\xi=\rho/a$ so $0 \le \xi \le 1$. Thus

$$\psi_v = -\pi B_0 a^2 \xi^2 \left[\mu_0 + (\mu_a - \mu_0)\xi^2/2\right]. \tag{3.38}$$

At the plasma boundary

$$\psi_{va} = \psi_v(\xi=1) = -\pi B_0 a^2(\mu_a + \mu_0)/2. \tag{3.39}$$

Now we use (3.33) to determine λ_1

$$\lambda_1 = -\ln\left(\frac{p_0}{p_a}\right)\left(\frac{\pi B_0 a^2(\mu_a + \mu_0)}{2}\right)^{-1}. \tag{3.40}$$

Substituting (3.40) into (3.30), we obtain an expression for the canonical pressure profile

$$p_c(\xi) = p_0 \exp\left\{-2\xi^2 \frac{\ln(p_0/p_a)}{1+\mu_a/\mu_0}\left[1+\frac{\xi^2}{2}\left(\frac{\mu_a}{\mu_0}-1\right)\right]\right\}. \tag{3.41}$$

Unlike the tokamak, where the canonical pressure profile is the power function of radial coordinates (2.18) in the stellarator the dependence on radial coordinate is

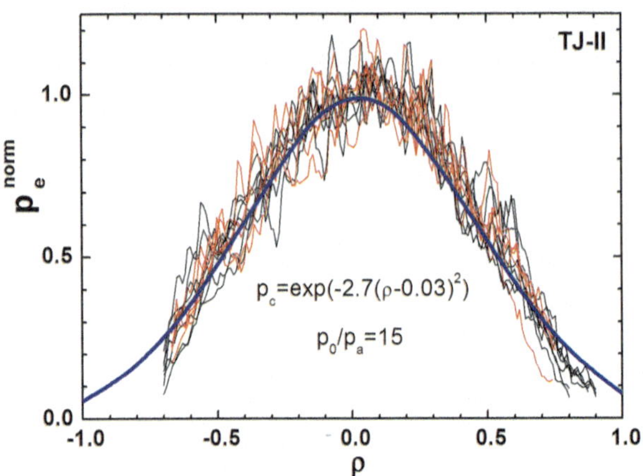

Fig. 3.1 Normalized experimental profiles of electron pressure for many discharges of TJ-II device with on- and off-axis ECR heating (*black* and *red fine lines*) at different time moments and the normalized canonical pressure profile $p_c(\psi)/p_0 = \exp\left[-2.7(\xi-0.03)^2\right]$. at $p_0/p_a = 15$ (*thick blue line*)

exponential one (3.41). This difference is due to the fact that the structure of the tokamak magnetic surfaces is defined by plasma current, but in the stellarator (in the selected approximation) by helical currents, located outside the plasma.

Stellarator TJ-II relates to devices with a very small shear ($\mu_a/\mu_0 \sim 1$). In this case, the canonical pressure profile (3.41) is described by a simple Gaussian. In Fig. 3.1, in addition to the experimental profiles of the electron pressure [4] a canonical pressure profile is given at $p_0 / p_a = 15$:

$$p_c(\psi)/p_0 = \exp\left[-2.7(\xi-0.03)^2\right]. \tag{3.42}$$

The correction in the index is due to a slight shift of plasma from the center of the chamber. It can be seen that the canonical profile is close to the experimental pressure profiles. Note that the representation of pressure profiles in exponential form (3.41) has recently been used in the analysis of experimental profiles of the pressure in LHD device [5].

In conclusion of this chapter, we note the following. For a long time it was believed that the transport of energy and particles in stellarators is determined by the neoclassical transport model and, therefore, the stellarator is radically different to the tokamak, where the transport is anomalous and is determined by plasma turbulence. However, gradually, the experimental facts cast doubt over this paradigm. Observation of the H-mode and internal transport barriers, the proximity of the stellarator scaling for the energy confinement time to scaling of the tokamak, all of these features led to the idea that turbulent transport in stellarators also plays an important role. In this chapter, we show that in stellarators the existence of self-consistent pressure profiles is also possible. This is another argument in support to the prominent role of the anomalous transport.

Note that in tokamaks the temperature profiles of electrons and ions are also self-consistent, which is not shown so far for the stellarator. Apparently, this difference is determined by the absence or smallness of the total current in stellarators and, consequently, the limited impact of Ohm's law. As a result, the temperature profiles in the stellarator easily change their shape but in tokamaks such change is more difficult. The existence of two self-consistent profiles in tokamaks (for temperature and pressure) allows one to build on this base a transport model concerning to the temperature of ions and electrons and to the density of plasma, as set out in the following chapters. Unfortunately one cannot see now the second self-consistent profile for the stellarator, so the problem of creating of a complete transport model based on the existence of canonical profiles for a stellarator is still open.

References

1. Dnestrovskij, Yu.N., Melnikov, A.V., Pustovitov, V.D.: Approach to canonical pressure profiles in stellarators. Plasma Phys. Control. Fusion. **51**, 015010 (2009)
2. Hsu, J.Y., Chu, M.S.: The tokamak equilibrium profile. Phys. Fluids. **30**, 1221 (1987)
3. Pustovitov, V.D.: Theoretical principles of the plasma-equilibrium control in stellarators. In: Kadomtsev, B.B., Shafranov, V.D. (eds.) Reviews of Plasma Physics, vol. 21, p. 1. Consultants Bureau, New York (2000)
4. Melnikov, A.V., Eliseev, L.G., Pastor, I., et al.: Pressure profile shape constancy in L-mode stellarator plasmas. In: Proceedings of the 34-th EPS Conference on Plasma Phys. and Controlled Fusion, Warsaw, Poland, ECA, vol. 31F, Rep. P-2.060 (2007)
5. Miyazawa, J., Yamada, H., et al.: Electron pressure profiles in high-density neutral beam heated plasmas in the large helical device. J. Plasma Fusion Res. **81**, 302 (2005)

Chapter 4
Theoretical Limitations for Scaling Laws and Transport Coefficients

Abstract In this chapter theoretical limits on plasma scaling laws and transport coefficients are derived. General properties of multi-machine and one-machine scaling laws for the energy confinement time τ_E are discussed. Possible reasons for the discrepancies between different scaling laws are examined. We find that plasma self-organization is the main reason for the different dependencies of the ITER confinement time scaling on the toroidal and poloidal magnetic field. The Taylor-Connor scale invariance principle is introduced and its consequences on the interpretation of scaling laws and limitations of the parametric form of transport coefficients are discussed.

4.1 Plasma Energy Confinement Time and Scaling Laws

Before we develop a transport model, we turn to the restrictions, which the theory imposes on the transport coefficients and the so-called scaling laws. Let's clarify some definitions. The energy confinement time τ_E during a stationary phase of the discharge is defined by the expression

$$\tau_E = \frac{W}{P}, \tag{4.1}$$

where W is the plasma thermal energy storage,

$$W = \frac{3}{2}\int_V n(T_e + T_i)dV, \tag{4.2}$$

P is the total power absorbed by plasma, n is the plasma density, T_e and T_i are the temperatures of the electrons and ions, V is the plasma volume, $P = P_{OH} + P_{aux} - P_{rad}$, P_{OH} is the Ohmic heating power, P_{aux} is the power from external heat sources, P_{rad} is the radiated power.

The dependence of the energy confinement time τ_E on the plasma geometry and engineering parameters is usually called a scaling law. Until the end of the eighties scaling laws were determined for each device individually, e.g. the "Alcator scaling", the "T-11 scaling", etc. However, in connection with work on the ITER reactor project, an inter-machine database was created in the early nineties. The "ITER

Yu.N. Dnestrovskij, *Self-Organization of Hot Plasmas*, DOI 10.1007/978-3-319-06802-2_4,

scaling laws" were determined using this database containing the results of experiments from many devices. These scaling laws for various plasma modes are being continuously improved. They are used to predict the operational point of ITER and other devices in design [1–5]. The global ITER database is described in [6].

4.2 Interplays Between Parameters Describing the Plasma State

We assume that the general scaling of the energy confinement time τ_E contains six significant independent variables. It could be the "engineering" variables (the plasma density n, deposited power P, the toroidal magnetic field B, the minor radius a, the total current of plasma I, the aspect ratio A), "physical" variables (n, the average temperature T, B, a, I, A) or "dimensionless" variables (ρ^*, ν^*, β, d^*, q, A). The database provides for each discharge six of these parameters and the value of energy confinement time τ_E. The scaling law expression (or, shortly, "scaling") can thus be written as follows:

$$\tau_E = f_0 n^{c1} P^{c2} B^{c3} a^{c4} I^{c5} A^{c6}, \tag{4.3}$$

$$B\tau_E = f_1 n^{b1} T^{b2} B^{b3} a^{b4} I^{b5} A^{b6}, \tag{4.4}$$

$$B\tau_E = f_2 (\rho^*)^{\alpha 1} (\nu^*)^{\alpha 2} \beta^{\alpha 3} (d^*)^{\alpha 4} q^{\alpha 5} A^{\alpha 6}, \tag{4.5}$$

where f_0, f_1 and f_2 can be functions of other dimensional or dimensionless parameters. Below we shall discuss, why in the left hand side of formulas (4.4) and (4.5) the product $B\tau_E$ is used, rather than the energy confinement time τ_E.

The dimensionless parameters are defined as follows:

$$\rho^* = \frac{r_L}{a} = \frac{\upsilon_T}{\omega_B a}, \quad \nu^* = \frac{\nu_{ii}}{\nu_B}, \quad \beta = \frac{8\pi n T}{B^2},$$

$$d^* = \frac{\upsilon_T}{\omega_0 a}, \quad q = \frac{5a^2 B}{IR}, \quad A = \frac{R}{a} \tag{4.6}$$

where

$$\omega_B = \frac{eB}{mc}, \quad \nu_{ii} = \frac{4\pi^{1/2} e^4 nL}{3m^{1/2} T^{3/2}}, \quad \nu_B = \frac{\varepsilon^{3/2} \upsilon_T}{2^{1/2} qR},$$

$$\omega_0 = \left(\frac{4\pi n e^2}{m}\right)^{1/2}, \quad \varepsilon = \frac{1}{A}, \tag{4.7}$$

where υ_T is the thermal velocity. After omitting numerical factors and universal constants, we obtain:

$$\rho^* \propto \frac{T^{1/2}}{aB}, \quad \nu^* \propto \frac{na^2BA^{3/2}}{IT^2}, \quad \beta \propto \frac{nT}{B^2},$$

$$d^* \propto \frac{T^{1/2}}{an^{1/2}}, \quad q \propto \frac{aB}{IA}. \tag{4.8}$$

The six variables introduced in (4.8) are called essential. We also give the inverse relationships:

$$a \propto \frac{(\rho^*)^2 qA^{5/2}}{\nu^*(d^*)^4\beta}, \quad B \propto \frac{\nu^*(d^*)^5\beta^{3/2}}{q(\rho^*)^4 A^{5/2}}, \quad T \propto \frac{(d^*)^2\beta}{(\rho^*)^2},$$

$$n \propto \frac{\beta^3(\nu^*)^2(d^*)^8}{q^2(\rho^*)^6 A^5}, \quad I \propto \frac{\beta^{1/2}d^*}{qA(\rho^*)^2}. \tag{4.9}$$

Substituting (4.8) and (4.9) in the scaling laws (4.4, 4.5), one can obtain a general formula for the direct and inverse transformation of six-dimensional vectors $\boldsymbol{\alpha}=\{\alpha_1, \alpha_2, \alpha_3, \alpha_4, \alpha_5, \alpha_6\}$ and $\mathbf{b}=\{b_1, b_2, b_3, b_4, b_5, b_6\}$,

$$\boldsymbol{\alpha} = \mathbf{B}_6\mathbf{b}, \quad \mathbf{b} = \mathbf{B}_6^{-1}\boldsymbol{\alpha}, \tag{4.10}$$

where $\mathbf{B}_6$ is a linear transformation. Unfolded, the direct and inverse transformations $\mathbf{B}_6$ and $\mathbf{B}_6^{-1}$ are as follows

$$\begin{aligned}
\alpha_1 &= -6b_1 - 2b_2 - 4b_3 + 2b_4 - 2b_5,\\
\alpha_2 &= 2b_1 + b_3 - b_4,\\
\alpha_3 &= -3b_1 + b_2 + \frac{3}{2}b_3 - b_4 + \frac{1}{2}b_5,\\
\alpha_4 &= 8b_1 + 2b_2 + 5b_3 - 4b_4 + b_5,\\
\alpha_5 &= -2b_1 - b_3 + b_4 - b_5,\\
\alpha_6 &= -5b_1 - \frac{5}{2}b_3 + \frac{5}{2}b_4 - b_5 + b_6,
\end{aligned} \tag{4.11}$$

$$\begin{aligned}
b_1 &= \alpha_2 + \alpha_3 - \frac{1}{2}\alpha_4,\\
b_2 &= \frac{1}{2}\alpha_1 - 2\alpha_2 + \alpha_3 + \frac{1}{2}\alpha_4,\\
b_3 &= -\alpha_1 + \alpha_2 - 2\alpha_3 + \alpha_5,\\
b_4 &= -\alpha_1 + 2\alpha_2 - \alpha_4 + \alpha_5,\\
b_5 &= -\alpha_2 - \alpha_5,\\
b_6 &= \frac{3}{2}\alpha_2 - \alpha_5 + \alpha_6.
\end{aligned} \tag{4.12}$$

The determinant of the matrix $\mathbf{B}_6$ is equal to −2; i. e. the transformations (4.11–4.12) are non-degenerate, so the chosen system of dimensionless parameters is independent. The transition from engineering to physical variables is given by

$$\tau_E \propto \frac{nTa^3 A}{P}$$

or

$$P \propto \frac{nTa^3 A}{\tau_E}. \tag{4.13}$$

Substituting (4.13) for P or T in the scalings (4.3, 4.4), we find a mutual conversion of six-dimensional vectors **b** and $\mathbf{c}=\{c_1, c_2, c_3, c_4, c_5, c_6\}$. Due to the fact that the transformation formula (4.13) depends on τ_E, the transformations

$$\mathbf{b} = \mathbf{D}\cdot\mathbf{c}, \quad \mathbf{c} = \mathbf{D}^{-1}\cdot\mathbf{b} \tag{4.14}$$

are fractional-linear:

$$b_1 = \frac{c_1 + c_2}{1+c_2}, \quad b_2 = \frac{c_2}{1+c_2}, \quad b_3 = \frac{1+c_2+c_3}{1+c_2},$$

$$b_4 = \frac{3c_2 + c_4}{1+c_2}, \quad b_5 = \frac{c_5}{1+c_2}, \quad b_6 = \frac{c_6 + c_2}{1+c_2}, \tag{4.15}$$

$$c_1 = \frac{b_1 - b_2}{1-b_2}, \quad c_2 = \frac{b_2}{1-b_2}, \quad c_3 = \frac{b_3 - 1}{1-b_2},$$

$$c_4 = \frac{b_4 - 3b_2}{1-b_2}, \quad c_5 = \frac{b_5}{1-b_2}, \quad c_6 = \frac{b_6 - b_2}{1-b_2}. \tag{4.16}$$

In order to convert the scaling from engineering to dimensionless variables, one has sequentially to transform the scaling (4.3) to (4.4) by the transformation (4.15), and then scaling (4.4) to (4.5) by transformation (4.11).

Remark In [2, 7] the refined formulas for parameter q and the value of τ_E are used instead of (4.6, 4.8) and (4.13)

$$q \propto \frac{akB}{IA}, \quad \tau_E \propto \frac{nTa^3 kA}{P},$$

where k is the elongation of the plasma cross section. As a result, k becomes an essential variable, and the total number of independent variables increases to seven resulting in the somewhat more complex formulas of type (4.11−4.13) and (4.15−4.16). However, this does not change the physical meaning of the scaling, and therefore we prefer to use only six independent variables. The reader can easily derive, if necessary, seven-dimensional analogues of (4.11−4.12) and (4.15–4.16), using the above described algorithms.

4.3 One-Machine Scaling

Since the minor radius a and aspect ratio A for a given device are fixed, there are only four independent variables in the engineering and physical variables (n, P, B, I) and (n, T, B, I). The scalings in this case can be written as follows

$$\tau_E = f_1 n^{c1} P^{c2} B^{c3} I^{c5} \quad \text{and} \quad B\tau_E = f_0 n^{b1} T^{b2} B^{b3} I^{b5}, \tag{4.17}$$

and the transitions from one set of variables to the other one are defined by relations (4.13), in which the variables a and A are omitted

$$\tau_E \propto \frac{nT}{P} \quad \text{or} \quad P \propto \frac{nT}{\tau_E}. \tag{4.18}$$

After substitution of (4.18) alternately in the first and second expression for the scaling law (4.17), we find for the mutual transformation of exponents of the parameters in the scaling, using the four-dimensional vectors $\mathbf{c}^4 = \{c_1, c_2, c_3, c_5\}$ and $\mathbf{b}^4 = \{b_1, b_2, b_3, b_5\}$, the following relation:

$$\mathbf{b}^4 = \mathbf{D}\cdot\mathbf{c}^4, \quad \mathbf{c}^4 = \mathbf{D}^{-1}\cdot\mathbf{b}^4 \tag{4.19}$$

The four-dimensional transformation $\mathbf{D}$ and $\mathbf{D}^{-1}$ turns out to be fractional- linear, as was the case for (4.15–4.16):

$$b_1 = \frac{c_1 + c_2}{1 + c_2}, \quad b_2 = \frac{c_2}{1 + c_1},$$

$$b_3 = \frac{1 + c_2 + c_3}{1 + c_2}, \quad b_5 = \frac{c_5}{1 + c_2} \tag{4.20}$$

$$c_1 = \frac{b_1 - b_2}{1 - b_2}, \quad c_2 = \frac{b_2}{1 - b_2},$$

$$c_3 = \frac{b_3 - 1}{1 - b_2}, \quad c_5 = \frac{b_5}{1 - b_2}. \tag{4.21}$$

These transformations are similar to the general formulas (4.15–4.16), but they include only four variables. Note that these four physical variables translate into five dimensionless variables $(\rho^*, \nu^*, \beta, d^*, q)$. This implies that the dimensionless variables in the one-machine scaling ($a = \text{const}$, $A = \text{const}$), are not independent.

Formulas for the transformation of the physical variables to the dimensionless ones, according to (4.8), are:

$$\rho^* \propto \frac{T^{1/2}}{B}, \quad \nu^* \propto \frac{nB}{T^2 I}, \quad \beta \propto \frac{nT}{B^2}, \quad q \propto \frac{B}{I}, \quad d^* \propto \frac{T^{1/2}}{n^{1/2}}. \tag{4.22}$$

From (4.22) it is easy to verify the relation

$$(d^*)^4 \propto \frac{(\rho^*)^2 q}{\nu^* \beta}. \tag{4.23}$$

It follows that one of the variables can be excluded from the scaling. Back in 1975, B.B. Kadomtsev pointed out [3] that the energy confinement time should not depend on the Debye length, so it is natural to exclude the variable d^* from the scaling. In this case, the scaling in dimensionless variables has the form:

$$B\tau_E = f_0(\rho^*)^{\alpha 1}(\nu^*)^{\alpha 2}\beta^{\alpha 3}q^{\alpha 5} \tag{4.24}$$

After substitution of (4.23) into (4.9) and dropping a minor radius a and an aspect ratio A, we obtain the formulas for the inverse transformation of the dimensionless variables into the physical ones:

$$n \propto \frac{\beta}{(\rho^*)^2}, \quad T \propto \frac{\beta^{1/2} q^{1/2}}{(\nu^*)^{1/2} \rho^*}, \quad B \propto \frac{\beta^{1/4} q^{1/4}}{(\nu^*)^{1/4} (\rho^*)^{3/2}}, \quad I \propto \frac{\beta^{1/4}}{(\rho^*)^{3/2} q^{3/4} (\nu^*)^{1/4}}. \tag{4.25}$$

Due to the multiplicativity of the relations (4.22, 4.25), the vectors $\boldsymbol{\alpha}^4 = \{\alpha_1, \alpha_2, \alpha_3, \alpha_5\}$ and $\mathbf{b}^4 = \{b_1, b_2, b_3, b_5\}$ are linked by non-degenerated, direct and inverse linear transformations

$$\boldsymbol{\alpha}^4 = \mathbf{B}_4 \cdot \mathbf{b}^4, \quad \mathbf{b}^4 = \mathbf{B}_4^{-1} \cdot \boldsymbol{\alpha}^4 \tag{4.26}$$

$$\begin{aligned}
\alpha_1 &= -2b_1 - b_2 - \frac{3}{2}b_3 - \frac{3}{2}b_5, \\
\alpha_2 &= -\frac{1}{2}b_2 - \frac{1}{4}b_3 - \frac{1}{4}b_5, \\
\alpha_3 &= b_1 + \frac{1}{2}b_2 + \frac{1}{4}b_3 + \frac{1}{4}b_5, \\
\alpha_5 &= \frac{1}{2}b_2 + \frac{1}{4}b_3 - \frac{3}{4}b_5,
\end{aligned} \tag{4.27}$$

$$\begin{aligned}
b_1 &= \alpha_2 + \alpha_3, \\
b_2 &= \frac{1}{2}\alpha_1 - 2\alpha_2 + \alpha_3, \\
b_3 &= -\alpha_1 + \alpha_2 - 2\alpha_3 + \alpha_5, \\
b_5 &= -\alpha_2 - \alpha_5,
\end{aligned} \tag{4.28}$$

These formulas are not a simple consequence of the general formulas (4.11–4.12), since in their derivation the relation (4.23) was used. Also for the transformations (4.27–4.28) we find:

$$\det \mathbf{B} = \frac{1}{2}, \quad \det \mathbf{B}^{-1} = 2. \tag{4.29}$$

We emphasize that we have put $\alpha_4 = 0$ without any additional assumption. In a one-machine scaling we always have only four independent variables.

4.4 Multi-Machine Scaling

For the sake of brevity in this Section the multi-machine database will be called an "ensemble" and the one-machine database a "cluster." The ensemble is a set of clusters. Each cluster contains four independent variables and is characterized by two fixed variables a and A. Although the ensemble contains six variables, the variation of two of them, a and A, is much smaller than of the four other ones. Indeed, in each cluster, a and A are fixed, so the paired correlations of the other variables with the variables a and A are zero and this is valid over the entire ensemble. At the same time it is obvious that the averaged values over a cluster of variables such as current and power ($<I>$ and $<P>$), increase with increasing values for a. The larger a, the greater are $<I>$ and $<P>$. We call this property for pairs of variables like e.g. ($<I>$, a) and ($<P>$, a) as "collinearity". Collinearity of variables means that the ensemble has no current and power values far away from the "ratio of collinearity." A simple comparison of the currents, powers and minor radius in clusters shows that the ratio of collinearity for current and power is given approximately by $<I> \propto a^2$, $<P> \propto a^2$. Figure 4.1 illustrates this for the current. Here the plasma minor radius a is plotted along the horizontal axis, and the value of $(I/3)^{1/2}$ along the vertical one. Straight diagonal lines correspond to the relationships $\xi = I[\mathrm{MA}]/a^2[\mathrm{m}^2] = 3$ and 1.5. We also indicate the zones in this diagram for which there are no experimental data.

Another important feature of the multi-machine database is the small number of clusters. It is much smaller than the number of discharges in each cluster. As an example, the general ensemble for the construction of the ITER scaling consists of 10–12 clusters, and the number of discharges in the clusters ranges from a few dozen to two or three hundreds. This means that there are only 10–12 different values for the pair of variables (a, A), characteristic for each individual cluster. Under such conditions, the separation of the exponents c_4 and c_6 in the scaling (4.3) is an ill-posed problem. Collinearity of variables signifies that in a multi-machine scaling the dependence of τ_E on current and power may not be quite the same as in one-machine scaling, where these variables are independent, and values for the minor radius a, are absent.

How, then, are we to understand the multi-machine scaling?

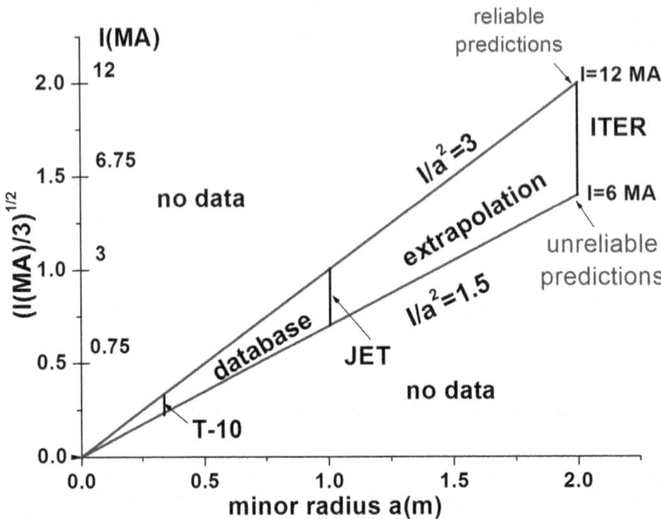

Fig. 4.1 The collinearity of current I with minor radius a in a multi-machine database

Suppose that in the database the variables I and P are collinear with the minor radius a, as follows: $I \propto a^2$, $P \propto a^2$. For many devices giving the main contribution to the ITER database, the ratios of collinearity are close to $\xi_I = I/a^2 \sim 3$, $\xi_P = P/a^2 \sim 20$, with I in MA, a in m and P in MW. Therefore, a scaling of the type (4.3) can only give a reasonable prediction for discharges on existing or projected devices, for which $\xi_I \sim 3$, $\xi_P \sim 20$. For ITER $a = 2$ m, so the scaling of the type (4.3) would make a reasonable prediction for the energy confinement time for discharges with a current of the order of 12 MA and a heating power of 80 MW. Scaling (4.3) cannot be applied to discharges with a heating power, for example, of $P = 160$ MW, and a current of $I = 6$ MA. Predictions obtained in that case will be evidently unreliable

For a more accurate interpretation of the meaning of a multi-machine scaling one can proceed as follows. Suppose that, as already mentioned, the multi-machine database (ensemble) is composed of separate databases from different devices. For each database (cluster), we introduce variables $\xi_I = I/a^2$ and $\xi_P = P/a^2$, and determine their average values $<\xi_I>$, $<\xi_P>$ and dispersions $\delta_I = \left\langle (\xi_I - <\xi_I>)^2 \right\rangle^{1/2}$, $\delta_p = \left\langle (\xi_p - <\xi_p>)^2 \right\rangle^{1/2}$. In the full database (i.e. in the ensemble), these values are functions of the size a. Let us extrapolate the average data and dispersions on the size of the device. Then the multi-machine scaling will give the most reliable prediction for those discharges of the projected device, for which the parameters ξ_I and ξ_P coincide with the extrapolated values of $<\xi_I>$ and $<\xi_P>$. The extrapolated values for the dispersions will determine the range of variation for the discharge parameters of the device, for which the scaling gives reasonable results. Of course, the extrapolation is an ill-posed problem; therefore one needs in each case to find a reasonable regularizer. For example, one can to provide the extrapolation using polynomials of the first or second order.

4.5 Examples

4.5.1 Multi-Machine Scaling ITER

The most popular scaling for the H-mode, found using the multi-machine database is [2, 4]:

$$\tau_E^H(\text{ITER}-98(\text{y},2)) = 0.0562 n^{0.41} P^{-0.69} B^{0.15} a^{1.97} I^{0.93} A^{1.39} k^{0.78} M^{0.19}, \tag{4.30}$$

where M is the ratio of the ion mass to the proton mass and k is the elongation of the plasma cross-section. According to (4.15–4.16) and (4.11–4.12), we find in physical and dimensionless variables (omitting parameters k and M)

$$B\tau_E^H(\text{ITER}-98(\text{y},2)) = f_0 n^{-0.9} T^{-2.22} B^{1.48} a^{-0.32} I^3 A^{2.25}, \tag{4.31}$$

$$B\tau_E^H(\text{ITER}-98(\text{y},2)) = f_0 (\rho^*)^{-2.73} (\nu^*)^{0.01} \beta^{-0.88} (d^*)^0 q^{-3} A^{-0.74}. \tag{4.32}$$

At first glance the exponents in formula (4.31) have no physical meaning. According to this scaling τ_E decreases with increasing density; τ_E even decreases with increasing minor radius a. However, in scaling (4.31), all the variables are independent. Therefore, when changing one parameter, we have to maintain the other ones unchanged. However, with increasing density in the experiment the temperature falls, and to keep it the same, we must increase the heating power. In this process, by virtue of (4.30), the value of τ_E diminishes and that is reflected by the negative density exponent in (4.31). When increasing the plasma current the energy confinement is improved, but not as much as implied by the exponent of the current ($b_5=3$, while $c_5=0.93$). However, when confinement is improved, the temperature rises, but its exponent in (4.31) is negative. To maintain the temperature it is necessary to reduce the deposited power and this in turn also increases τ_E. This "synergetic" effect is reflected by the very large value of the exponent $b_5=3$.

The dependence of τ_E on the normalized Debye length d^* has disappeared in (4.32). It is made artificially but the possibility of such an effect is determined by the collinearity of the values of plasma current I and deposited power P with the minor radius a in the multi-machine database. Due to the strong correlation between the variables, the problem of determining the exponents of plasma current, heating power, and minor radius is ill-posed and requires regularization. One method is to adopt the regularization assumption that $\alpha_4=0$. J. Connor has shown [3] that this assumption is equivalent to assuming plasma quasi-neutrality. We will discuss this issue below. With the transformation formulas (4.12) and (4.16), this condition is converted into:

$$5 + 8c_1 + 3c_2 + 5c_3 - 4c_4 + c_5 = 0. \tag{4.33}$$

Thus, of the first five parameters in formula (4.30) only four are independent, and the problem is regularized. It is easy to verify that the exponents of scaling (4.30) satisfy (4.33). From (4.32) we also see that the dependence of the scaling on the normalized collision frequency ν^* has nearly disappeared, i.e. the exponents in the scaling (4.30) also satisfy the relation

$$|\alpha_2| \ll 1 \text{ or } |1+2c_1+c_3-c_4| \ll 1|. \tag{4.34}$$

In expression (4.34) the exponents c_2 and c_5 associated with the variables P and I are absent. This reflects the independence of the energy confinement time on the collision frequency, as found experimentally. We will show that this effect corresponds to the kinetic description of plasma without collisional term [5].

A careful inspection of scaling (4.30) reveals the small value of the exponent of the toroidal magnetic field (c_3=0.15) and a nearly 6 times larger value of the exponent of the plasma current (c_5=0.93). What would be the reason for such a difference?

It is known that the width of the pressure profile depends on the value of q at the plasma edge (this follows from the principle of self-consistency of the pressure profile, see Introduction). With increasing values of q the pressure profile width decreases and thus the plasma stored energy also decreases (keeping the other engineering parameters fixed). Since $q \propto B/I$ the effect of the plasma current increase on the width of the pressure profile is opposite to the effect of the toroidal magnetic field increase. If the scaling would reflect only the influence of the shape of the pressure profile, the exponents of the plasma current and magnetic field would have opposite signs. However, in addition to the profile shape, the plasma current and magnetic field also determine heat transport in a similar way: the higher the plasma current and the magnetic field, the lower the transport. This is e.g. shown in the neo-classical plateau regime, where the ion heat diffusivity $\chi_i \propto (IB)^{-1}$. Thus, when increasing the plasma current the positive effect on energy confinement from both the broadening of the pressure profile and reducing the transport is added up, while when increasing the magnetic field, the negative effect from narrowing the pressure profile and the positive effect of reducing plasma transport partially compensate each other. Therefore, the exponents c_3 and c_5 in scaling (4.30) are very different. This difference implicitly confirms the existence and importance of the effect of pressure profile self-consistency.

4.5.2 Experimental One-Machine Scaling Laws

Let us consider, as an example, the one-machine scaling derived in [7] for JET, using the JET database containing data from more than 300 H-mode discharges:

$$\tau_E \propto n^{0.41} I^{0.76} B^{0.26} P^{-0.4}. \tag{4.35}$$

Using (4.15) and (4.11), we can transform this scaling to physical and dimensionless variables. Scaling laws in engineering variables similar to (4.35) were also determined for other tokamaks. Attempts were also made at determining one-machine scalings directly in terms of dimensionless variables.

This can be done as follows. Let the scaling has the form (4.24)

$$B\tau_E = f_0(\rho^*)^{\alpha 1}(\nu^*)^{\alpha 2}\beta^{\alpha 3}q^{\alpha 5}$$

and suppose we want to determine experimentally the exponent α_1. Then for a selected series (usually a pair) of discharges the following conditions must be satisfied:

$$\nu^* = \text{const}, \quad \beta = \text{const}, \quad q = \text{const}. \tag{4.36}$$

These three conditions create a link between the four physical parameters *n*, *T*, *B*, *I*, so three of them can be expressed in terms of the fourth remaining one. Following (4.22), we have:

$$n \propto B^{4/3}, \quad T \propto B^{2/3}, \quad I \propto B, \tag{4.37}$$

Moreover

$$\rho^* \propto B^{-2/3}. \tag{4.38}$$

Suppose that for a pair of discharges with the magnetic fields B_1 and B_2 the conditions (4.36–4.38) are satisfied. Then

$$\frac{B_1\tau_{E1}}{B_2\tau_{E2}} = \left(\frac{\rho_1^*}{\rho_2^*}\right)^{\alpha 1} = \left(\frac{B_1}{B_2}\right)^{-2/3\alpha 1}. \tag{4.39}$$

So, we obtain

$$\alpha_1 = \frac{3}{2}\ln\left(\frac{B_1\tau_{E1}}{B_2\tau_{E2}}\right)\left[\ln\left(\frac{B_2}{B_1}\right)\right]^{-1}. \tag{4.40}$$

One can proceed in similar way to identify the other exponents of the scaling. Thus to determine α_2, α_3 and α_5 the parameters of the pair of discharges must satisfy the conditions:

$$\begin{aligned} &n \propto \text{const}, \quad T \propto B^2, \quad I \propto B, \\ &n \propto B^4, \quad T \propto B^2, \quad I \propto B, \\ &n \propto \text{const}, \quad T \propto B^2, \quad I \propto B^{-3}, \end{aligned} \tag{4.41}$$

Table 4.1 Exponents in multi-machine dimensionless ITER scaling and in one-machine dimensionless scalings

Exponents	ITER-98(y,2)	DIII-D	JET [2, 4]	JET [6]	JT-60U	MAST	ASDEX-Upgrade	C-Mod
α_1 (ρ*)	−2.73	−2.7	−2.7	−3.4	−2.8		−3	−3.1
α_2 (ν*)	0.01	−0.35	−0.27	−0.33		−0.82		−1
α_3 (β)	−0.88	0.1	−0.05	0.35	0.65		0.85	
α_5 (q)	−3	−1.4		−0.93		−0.85		

One can easily find the formulas for the exponents:

$$\alpha_2 = \frac{1}{4}\ln\left(\frac{B_1\tau_{E1}}{B_2\tau_{E2}}\right)\left[\ln\left(\frac{B_2}{B_1}\right)\right]^{-1},$$

$$\alpha_3 = -\frac{1}{4}\ln\left(\frac{B_1\tau_{E1}}{B_2\tau_{E2}}\right)\left[\ln\left(\frac{B_2}{B_1}\right)\right]^{-1},$$

$$\alpha_5 = -\frac{1}{4}\ln\left(\frac{B_1\tau_{E1}}{B_2\tau_{E2}}\right)\left[\ln\left(\frac{B_2}{B_1}\right)\right]^{-1}. \tag{4.42}$$

Of course, satisfying the conditions (4.37) or (4.41) is not always easy experimentally. Therefore the accuracy with which the parameters α_k, are determined may be low.

The obvious question is, why one tries to determine experimentally the exponents in scalings expressed in dimensionless variables (ρ^*, ν^*, β, q), instead of in physical ones (n, T, B, I), and then transforming the scaling found using formulas (4.27)? It seems that the nature of variables in one-machine scaling and transformations (4.27–4.28) were not taken into account. The measurements of scaling in physical variables would be much simple. Indeed, for each pair of discharges one has to change only one physical variable, fixing the others, while using a scaling in dimensionless form one has to satisfy simultaneously the complicated conditions (4.37) or (4.41).

4.5.3 A Comparison of Scalings

A comparison of the multi-machine scaling ITER-98 (y, 2) with the experimental one-machine scalings from different devices is summarized in Table 4.1. The data in the second column for JET were obtained from the scaling (4.35) by using the conversion formulas (4.20) and (4.27). It is seen that only the exponent α_1 is approximately the same in all cases, while the other ones are very different. The difference between the exponents in the multi-machine and one-machine scalings is not unexpected. As discussed above, it is determined by the difference in the character of the databases: multi-machine experimental database contains highly correlated variables, while in one-machine database the variables are independent.

However, the exponents in one-machine scalings are also very different, and this has already been remarked in several studies [2, 4, 7–9]. The differences in the exponents of one-machine scalings can be understood from the specific properties of the individual devices (plasma geometry, first wall material, the amount of impurities in the plasma, the conditioning of the machine, etc.) and peculiarities of the discharges in the database (e.g. the presence of sawtooth and MHD modes). However, in our opinion, the main reason for the discrepancies is different: the used "basic" variables (for example, the engineering variables n, T, B, I) do not completely determine the energy confinement in the plasma. For H-mode discharges, the stored plasma energy depends significantly on the pedestal pressure. For ELMy H-mode discharges the time-averaged value of the pedestal depends on the type and nature of the ELMs (Edge Localized Modes) and the rate of pedestal recovery after the ELMs. The average value of the pressure pedestal and the rate of pedestal recovery both increase with increasing ELM period and this leads in turn to an increase in the stored plasma energy. Obviously, the nature of ELMs can be easily varied in plasmas with fixed values for the engineering variables, e.g. by changing the gas puffing in the discharge. Thus it becomes obvious that this effect only indirectly depends on the choice of the engineering variables used in the scaling. More details on the properties of the H-mode and ELMs will be discussed in Chap. 5.

4.6 Connor-Taylor Theory of Invariants

This section discusses the theoretical problem of how the set of independent variables is changed, when the set of equations describing the hot tokamak plasma changes? [5].

4.6.1 Transformation of the Complete System of Equations for the Plasma

We first consider the general case and assume that the distribution function of electrons and ions in a plasma f_k ($k=e, i$) is described by the Vlasov equation with the Landau collision term, and the electromagnetic fields by Maxwell's equations and the Poisson equation:

$$\frac{\partial f_k}{\partial t} + \mathbf{v}\nabla f_k + \frac{e_k}{m_k}(\mathbf{E} + \mathbf{v}\times\mathbf{B})\frac{\partial f_k}{\partial \mathbf{v}} = C(f, f), \tag{4.43}$$

$$\mathrm{rot}\mathbf{E} = -\frac{\partial \mathbf{B}}{\partial t}, \quad \mathrm{rot}\mathbf{B} = 4\pi\mathbf{j}, \tag{4.44}$$

$$\operatorname{div}\mathbf{E} = 4\pi\sum_k e_k \int f_k(\vec{x},\vec{\upsilon})d^3\upsilon, \tag{4.45}$$

where

$$\mathbf{j} = \sum_k \int e_k \vec{\upsilon} f_k(\vec{x},\vec{\upsilon})d^3\upsilon. \tag{4.46}$$

We define the following linear transformation of the independent and dependent variables

$$f \to \alpha f, \quad \mathbf{v} \to \theta\mathbf{v}, \quad \mathbf{x} \to \gamma\mathbf{x}, \quad \mathbf{B} \to \delta\mathbf{B}, \quad t \to \varepsilon t, \quad \mathbf{E} \to \eta\mathbf{E}, \tag{4.47}$$

leaving the equations (4.43–4.45) unchanged. Substituting (4.47) into (4.43–4.45), we obtain an algebraic system of 7 equations involving 6 parameters (α, θ, γ, δ, ε, η):

$$\frac{1}{\varepsilon} = \frac{\theta}{\gamma} = \frac{\eta}{\theta} = \delta = \alpha, \quad \frac{\eta}{\gamma} = \frac{\delta}{\varepsilon}, \quad \frac{\delta}{\gamma} = \alpha\theta^4, \quad \frac{\eta}{\gamma} = \alpha\theta^3. \tag{4.48}$$

The system (4.48) has no solution, so there is no transformation of the type (4.47), which leaves the system (4.43–4.45) invariable.

4.6.2 Approximation of Quasi-Neutral Plasma

Now consider the quasi-neutral plasma. To do this, replace the Poisson equation (4.45) by a simpler quasi-neutrality condition

$$\sum_k e_k \int f_k(\vec{x},\vec{\upsilon})d^3\upsilon = 0. \tag{4.49}$$

In this case, the last equation of (4.48) vanishes, and we obtain a truncated system of 6 equations

$$\frac{1}{\varepsilon} = \frac{\theta}{\gamma} = \frac{\eta}{\theta} = \delta = \alpha, \quad \frac{\eta}{\gamma} = \frac{\delta}{\varepsilon}, \quad \frac{\delta}{\gamma} = \alpha\theta^4. \tag{4.50}$$

This system has a unique solution that can be expressed in one parameter

$$\alpha = \theta^5, \quad \gamma = \theta^{-4}, \quad \delta = \theta^5, \quad \varepsilon = \theta^{-5}, \quad \eta = \theta^6. \tag{4.51}$$

We now determine transformations of integral variables. For the current density $\mathbf{j}$ (4.46) and the total current I we have

$$\mathbf{j} \rightarrow \alpha\theta^4 \mathbf{j} = \theta^9 \mathbf{j}, \tag{4.52}$$

$$I = \int_S j\, dS \rightarrow \theta I, \tag{4.53}$$

where S is the plasma cross section. Thus the total current is not an invariant of the transformation (4.47) for the system (4.43–4.44) in conjunction with the condition of quasi-neutrality. For the plasma density and temperature we have

$$n = \int f d^3 \upsilon \rightarrow \theta^8 n, \quad nT = \int \upsilon^2 f d^3 \upsilon \rightarrow \theta^{10} nT, \quad T \rightarrow \theta^2 T, \tag{4.54}$$

and for the heat flux

$$Q = \int \vec{\upsilon} \upsilon^2 f d^3 \upsilon \rightarrow \theta^{11} Q. \tag{4.55}$$

In steady state the heat flux integrated over the plasma boundary is equal to the absorbed power, thus for the energy confinement time (4.1) we can write:

$$\tau_E \sim \frac{nTa}{Q} \rightarrow \theta^{-5} \tau_E. \tag{4.56}$$

From (4.51) we see that $\varepsilon = \theta^{-5}$ so the product $B\tau_E$ transforms into itself:

$$B\tau_E \rightarrow B\tau_E, \tag{4.57}$$

and, therefore is an invariant of the transformation (4.47).

It is important to remark that the existence of a unique one-parametric solution of system (4.50) implies that from the six independent variables in the scaling (4.3), we can define five independent parameters that are invariant under the transformation (4.47)

$$J_1 = a^2 n, \quad J_2 = aT^2, \quad J_3 = a^5 B^4, \quad J_4 = IA/(aB) = 1/q, \quad J_5 = A. \tag{4.58}$$

Note that the dimensionless parameters ρ^*, v^*, β and d^* are also invariants of the transformation (4.47). This follows from the invariance of the right hand side of (4.8). Since any function of invariants is again invariant, many other invariants can be defined.

A general expression for the scaling of the invariant quantity $B\tau_E$, compatible with the system of equations for a quasi-neutral plasma (4.43, 4.44, 4.49) can be represented as an arbitrary function of the invariants (4.58), together with the "trivial" parameters k, δ, …:

$$B\tau_E = F(a^2 n, aT^2, a^5 B^4, IA/(aB), A, k, \delta \ldots). \tag{4.59}$$

If we choose for the arbitrary function F a product of power functions for the first five variables, then we have

$$B\tau_E = (a^2 n)^{a1}(aT^2)^{a2}(a^5 B^4)^{a3}(IA/(aB))^{a4} A^{a5} f(k,\delta...). \tag{4.60}$$

Using (4.4) and (4.60), we find the direct and inverse connections of a_k and b_k:

$$b_1 = a_1,\quad b_2 = 2a_2,\quad b_3 = 4a_3 - a_4,\quad b_4 = 2a_1 + a_2 + 5a_3 - a_4,$$
$$b_5 = a_4,\quad b_6 = a_4 + a_5, \tag{4.61}$$

$$a_1 = b_1,\quad a_2 = \frac{1}{2}b_2,\quad a_3 = \frac{1}{4}(b_3 + b_5),$$
$$a_4 = b_5,\quad a_5 = b_6 - b_5. \tag{4.62}$$

From (4.62) we see that the values of a_k do not depend on b_4, so the expression for b_4 in (4.61) expresses a compatibility condition between the representations (4.4) and (4.60) for quasi-neutral plasma. This can be transformed using (4.15–4.16) and (4.11–4.12), into:

$$2b_1 + \frac{1}{2}b_2 + \frac{5}{4}b_3 - b_4 + \frac{1}{4}b_5 = 0$$

or

$$\Delta \equiv 5 + 8c_1 + 3c_2 + 5c_3 - 4c_4 + c_5 = 0 \tag{4.63}$$

or

$$\alpha_4 = 0.$$

To quantify the proximity of the empirical scaling to the approximate physical model of a quasi-neutral plasma it is convenient to use the value

$$D = \frac{|\Delta|}{M}, \tag{4.64}$$

where $M = \max|c_k|$ is the absolute value of the maximum term in (4.63).

If we now substitute (4.9) and use (4.63) in a general expression of scaling (4.60), we obtain:

$$B\tau_E = f_2(\rho^*)^{\alpha 1}(\nu^*)^{\alpha 2}\beta^{\alpha 3}q^{\alpha 5}A^{\alpha 6}. \tag{4.65}$$

Using (4.63), the connection formulas (4.11–4.12) between α_k and b_k can be simplified as follows

$$\boldsymbol{\alpha} = \mathbf{C_q}\cdot\mathbf{b},\quad \mathbf{b}=\mathbf{C_q^{-1}}\cdot\boldsymbol{\alpha}, \tag{4.66}$$

$$\alpha_1 = -\left(2b_1 + b_2 + \frac{3}{2}b_3 + \frac{3}{2}b_5\right), \quad \alpha_2 = -\left(\frac{1}{2}b_1 + \frac{1}{4}b_2 + \frac{1}{4}b_5\right),$$

$$\alpha_3 = b_1 + \frac{1}{2}b_2 + \frac{1}{4}b_3 + \frac{1}{4}b_5, \quad \alpha_4 = 0,$$

$$\alpha_5 = \frac{1}{2}b_2 + \frac{1}{4}b_3 + \frac{1}{4}b_5, \quad \alpha_6 = \frac{5}{4}b_2 + \frac{5}{8}b_3 - \frac{3}{8}b_5 + b_6 \tag{4.67}$$

$$\begin{aligned} &b_1 = \alpha_2 + \alpha_3, \quad b_2 = \frac{1}{2}\alpha_1 - 2\alpha_2 + \alpha_3, \\ &b_3 = -\alpha_1 + \alpha_2 - 2\alpha_3 + \alpha_5, \quad b_4 = -\alpha_1 + 2\alpha_2 + \alpha_5, \\ &b_5 = -\alpha_2 - \alpha_5. \end{aligned} \tag{4.68}$$

Consider e.g. the ITER scaling for the H-mode (4.30) and for the L-mode

$$B\tau_E^L(\text{ITER} - 89\text{P}) = f_1 n^{0.1} P^{-0.5} B^{0.2} a^{1.5} I^{0.85} A^{1.2}. \tag{4.69}$$

In dimensionless variables, this scaling becomes:

$$B\tau_E^L(\text{ITER} - 89\text{P}) = f_2(\rho^*)^{-2.2}(\nu^*)^{-0.2}\beta^{-0.45}(d^*)^{0.3}q^{-1.5}A^{1.2}. \tag{4.70}$$

For scaling τ_E^H (ITER $-$98(y,2)) (as expressed in (4.30–4.32)) the exponent $\alpha_4 = 0$, therefore this scaling is consistent with the condition for a quasi-neutral plasma. For scaling (4.69–4.70) the value of exponent $\alpha_4 = 0.3 \neq 0$, and the representation (4.70) cannot be written as an expression of the form (4.65). However, if D is small, the condition (4.63) is neglected and the exponents α_k are calculated using formulas (4.67). The corresponding scalings will be marked with an asterisk. E.g. for scaling (4.69) we thus find:

$$B\tau_E^{L*}(\text{ITER} - 89\text{P}) = f_0(\rho^*)^{-2.05}(\nu^*)^{-0.27}\beta^{-0.53}q^{-1.43}A^{1.2}. \tag{4.71}$$

The scaling (4.71) can be referred as an "experimental scaling" for the L-mode (4.70), corrected by the quasi-neutrality condition. Note that the exponents in (4.71) do not significantly differ from the exponents in (4.70). This is due to the fact that the discrepancy D^L for the empirical scaling (4.69) is small: $D^L = 2.5\,\%$. This error is within the accuracy of the scaling, so it is possible to say that the empirical scaling (4.69) is not in contradiction with the condition for quasi-neutral plasma.

4.6.3 Approximation of a Collisionless Plasma

In this case, the system of equations describing the plasma is given by:

$$\frac{\partial f_k}{\partial t} + \boldsymbol{\upsilon}\nabla f_k + \frac{e_k}{m_k}(\mathbf{E} + \boldsymbol{\upsilon} \times \mathbf{B})\frac{\partial f_k}{\partial \boldsymbol{\upsilon}} = 0, \tag{4.72}$$

$$\mathrm{rot}\mathbf{E} = -\frac{\partial \mathbf{B}}{\partial t}, \quad \mathrm{rot}\mathbf{B} = 4\pi \mathbf{j}, \tag{4.73}$$

$$\mathrm{div}\mathbf{E} = 4\pi \sum_k e_k \int f_k(\vec{x}, \vec{v}) d^3 v. \tag{4.74}$$

Applying the transformation (4.47) to equations (4.72–4.74), we obtain the following system of six equations

$$\frac{1}{\varepsilon} = \frac{\theta}{\gamma} = \frac{\eta}{\theta} = \delta, \quad \frac{\eta}{\gamma} = \frac{\delta}{\varepsilon}, \quad \frac{\delta}{\gamma} = \alpha\theta^4, \quad \frac{\eta}{\gamma} = \alpha\theta^3. \tag{4.75}$$

The solution can be expressed with the parameter α, as follows:

$$\theta = 1, \quad \gamma = \alpha^{-1/2}, \quad \delta = \alpha^{1/2}, \quad \varepsilon = \alpha^{-1/2}, \quad \eta = \alpha^{1/2}. \tag{4.76}$$

The independent variables are transformed as follows:

$$n \to \alpha n, \quad T \to T, \quad B \to \alpha^{1/2} B, \quad a \to \alpha^{-1/2} a, \quad I \to I, \quad A \to A. \tag{4.77}$$

We have $\varepsilon\delta = 1$, so $B\tau_E$, as in the case of a quasi-neutral plasma, is an invariant of the transformation (4.47). The plasma current I is also invariant in this case. The required set of five invariants is:

$$J_1 = \frac{n}{B^2}, \quad J_2 = \frac{T}{a^2 B^2}, \quad J_3 = aB, \quad J_4 = I, \quad J_5 = A. \tag{4.78}$$

As before, we restrict ourselves to the power representation for the scaling

$$B\tau_E = f\left(\frac{n}{B^2}\right)^{d1}\left(\frac{T}{a^2 B^2}\right)^{d2}(ab)^{d3} I^{d4} A^{d5}. \tag{4.79}$$

Using (4.4) and (4.79), we find the direct and inverse relationships between b_k и d_k:

$$\begin{aligned} &b_1 = d_1, \quad b_2 = d_2, \quad b_3 = -2d_1 - 2d_2 + d_3, \\ &b_4 = -2d_2 + d_3, \quad b_5 = d_4, \quad b_6 = d_5, \end{aligned} \tag{4.80}$$

$$\begin{aligned} &d_1 = b_1, \quad d_2 = b_2, \quad d_3 = 2b_2 + b_4, \\ &d_4 = b_5, \quad d_5 = b_6. \end{aligned} \tag{4.81}$$

Hence the compatibility condition is as follows

$$2b_1 + b_3 - b_2 = 0,$$

or

$$\Delta \equiv 2c_1 + c_3 - c_4 + 1 = 0 \tag{4.82}$$

or

$$\alpha_2 = 0.$$

Using (4.82), the scaling in dimensionless variables becomes:

$$B\tau_E = f_0(\rho^*)^{\alpha 1} \beta^{\alpha 3} (d^*)^{\alpha 4} q^{\alpha 5} A^{\alpha 6}, \tag{4.83}$$

where

$$\boldsymbol{\alpha} = \mathbf{C}_v \cdot \mathbf{b}, \quad \mathbf{b} = \mathbf{C}_v^{-1} \cdot \boldsymbol{\alpha} \tag{4.84}$$

$$\alpha_1 = 2b_1 - 2b_2 - 2b_4 - 2b_5, \quad \alpha_2 = 0, \quad \alpha_3 = b_2 + \frac{1}{2}b_4 + \frac{1}{2}b_5,$$
$$\alpha_4 = -2b_1 + 2b_2 + b_4 + b_5, \quad \alpha_5 = -b_5, \quad \alpha_6 = -b_5 + b_6. \tag{4.85}$$

$$b_1 = \alpha_3 - \frac{1}{2}\alpha_4, \quad b_2 = \frac{1}{2}\alpha_1 + \alpha_3 + \frac{1}{2}\alpha_4, \quad b_3 = -\alpha_1 - 2\alpha_3 + \alpha_5,$$
$$b_4 = -\alpha_1 - \alpha_4 + \alpha_5, \quad b_5 = -\alpha_5, \quad b_6 = -\alpha_5 + \alpha_6. \tag{4.86}$$

To illustrate the above, let us consider again the scaling (4.69) for the L-mode. The discrepancy for this scaling is $D^L = 7\,\%$. Because of the moderately low value for this quantity we cannot unambiguously reject the model under consideration for the L-mode. Using relations (4.84–4.86), we find the dimensionless scaling

$$B\tau_E^{L*} = f(\rho^*)^{-3} \beta^{-0.15} (d^*)^{1.3} q^{-1.7} A^{0.7}. \tag{4.87}$$

We have already seen that for the H-mode scaling (4.30) $\alpha_2 = 0$ and $\alpha_4 = 0$, i.e. the compatibility conditions for both a quasi-neutral and collisionless plasma model are satisfied. Therefore, the dimensionless scaling in the case of a collisionless plasma is of the form (4.32) too. Thus, using the scaling expression (4.32) for H-mode, it is impossible to decide which of the models (4.43–4.45) or (4.72–4.74) describes best the H-mode.

4.6.4 Restrictions on the Transport Coefficients

For purely diffusive transport, for which the energy flux can be expressed as

$$Q = -n\chi \frac{\partial T}{\partial r}, \tag{4.88}$$

the heat diffusivity coefficient χ can be estimated as follows:

$$\chi = \frac{a^2}{\tau_E} = \frac{a^2 B}{F} = a^2 B F_1, \tag{4.89}$$

where F_1 is a function of the invariants of a particular model. The comparison of (4.89) with the empirical coefficients χ^{exp} will be held in the next Chap. 5.

References

1. http://tokamak-profiledb.ccfe.ac.uk/
2. Doyle E.J. et al. Plasma Confinement and transport. Nucl. Fusion 47, S18-S127 (2007) Sect. 5.3. Global scaling. P. S109
3. Kadomtsev, B.B.: Tokamaks and dimensional analysis. Sov. J. Plasma Phys. **1**, 295 (1975)
4. ITER Physics Basis, Chapter 2, Plasma confinement and transport. Nucl. Fusion **39**, 2201 (1999)
5. Connor, J.W., Taylor J.B.: Scaling laws for plasma confinement. Nucl. Fusion **17**, 1047 (1977)
6. McDonald, P.C., Cordey, J.G., Thomsen, K., et al.: Recent progress in the development and analysis of the ITPA global H-mode confinement database. Nucl. Fusion **47**, 147 (2007)
7. De Vries, P.C., et al.: Scaling of rotation and momentum confinement in JET plasmas. Nucl. Fusion **48**, 065006 (2008)
8. Petty, C.C., et al.: Nondimensional transport scaling in DIII-D: Bohm versus gyro-Bohm resolved. Phys. Plasma. **2**, 2342 (1995)
9. Valovich, M., et al.: Energy confinement and pellet fuelling in MAST. Proceeding of 23-rd International Conference on Fusion Energy 2010 (Daejeon, Korea, 2010) CD-ROM file EXC/P8-18. http://www-pub.iaea.org/MTCD/meetings/PDFplus/2010/cn180/cn180_papers/exc_p8-18.pdf

Chapter 5
Linear Version of the Canonical Profiles Transport Model (CPTM)

Abstract A linear version of the Canonical Profiles Transport Model (CPTM) is presented in this chapter. This version can be applied to Ohmically heated and L-mode plasmas in tokamaks. It is possible to apply this version also to the description of H-mode plasmas if the pedestal values of plasma temperatures and density are known. We present the expressions for the fluxes of heat, particles and toroidal rotation momentum, including critical gradients. These critical gradients are determined by the canonical profiles, defined in Chap. 2. The transport coefficients found by comparing the calculations with the experimental data are discussed. This is illustrated with examples from the tokamaks DIII-D, MAST and JET.

5.1 Transport Equations

To simplify the expressions, we assume in what follows that the vacuum toroidal magnetic field B_0 does not change in time. The set of transport equations describing the plasma density n, the temperature of electrons and ions, T_e, T_i, the potential of the poloidal magnetic field ψ and the density of the toroidal rotation momentum L in natural coordinates is given by [1]:

$$\frac{\partial n}{\partial t} + \mathrm{div}_\rho(G_1 \Gamma_n) = S_n, \tag{5.1}$$

$$\frac{3}{2}\frac{\partial n T_k}{\partial t} + \mathrm{div}_\rho(G_1 q_k) = P_k, \tag{5.2}$$

$$\sigma_\| \frac{\partial \psi}{\partial t} = \frac{1}{\mu_{00} B_0 \rho}\frac{\partial}{\partial \rho}\left(V' G \frac{\partial \psi}{\partial \rho}\right), \tag{5.3}$$

$$\frac{\partial L}{\partial t} + \mathrm{div}_\rho(G_1 q_L) = t_L. \tag{5.4}$$

Yu.N. Dnestrovskij, *Self-Organization of Hot Plasmas,* DOI 10.1007/978-3-319-06802-2_5,

The plasma equilibrium is determined by the solution of the Grad–Shafranov equation (2.2), and the coordinate ρ is defined according to (2.79). It is assumed in (5.4), that each magnetic surface rotates as a whole, therefore the angular frequency ω is a function of the magnetic surface, $\omega = \omega(\psi)$, the toroidal velocity is $\upsilon_t = \omega R$ and the density of the toroidal rotation momentum is given by the formula

$$L = nm_i \langle R\upsilon_t \rangle = nm_i \langle R^2 \rangle \omega, \tag{5.5}$$

In the formulas above R is the distance from the torus axis, m_i is the ion mass and t_L is the density of the external torque. Brackets $<\cdots>$ denote averaging over the magnetic surface. We assume that $\langle R^2 \rangle = R_0^2$ (where R_0 is a major radius of torus), and we neglect terms of the Shafranov shift in $(a/R_0)^2$ and higher order. In (5.1–5.4) Γ_n is the particle flux, q_k (k=e, i) are the electron and ion heat fluxes, q_L is the momentum flux, S_n is the particle source, P_k is the sum of heating power densities, radiated power density and energy exchange terms, $\sigma_{||}$ is the plasma longitudinal conductivity,

$$\mathrm{div}_\rho(.) = \frac{1}{V'}\frac{\partial}{\partial \rho}(V'(.)). \tag{5.6}$$

The metric coefficients V' and G are defined in (2.81), (2.85) of Chap. 2 and the metric coefficient G_1 equals to $G_1 = \langle (\nabla\rho)^2 \rangle$.

Equations (5.1) and (5.2) describe the conservation of the number of particles and stored energy in the plasma. Eq. (5.3) is Ohm's law for the longitudinal components of the plasma current. Equation (5.4) describes the conservation of toroidal angular momentum. To close the set (5.1)–(5.4) we need an expression relating the fluxes to the unknown functions n, T_k and L.

The correspondence between the equilibrium equation and Eq. (5.3) in the frame work of the ASTRA code is described in [1]. The differential Eqs. (5.1–5.4) should be supplemented by appropriate boundary and initial conditions.

5.2 Heat Flux, Particle Flux and Flux of Toroidal Momentum in Ohmic and *L* mode Plasmas

The expressions for the heat flux, particle flux and toriodal momentum flux are based on the following assumptions confirmed by the experimental facts:

a. In non-stationary Ohmic and L mode plasmas the profiles of electron and ion temperature, T_e, T_i, of the plasma pressure p=$n(T_e+T_i)$ and of the toroidal rotation velocity υ_t evolve to the canonical ones.
b. The temperature and pressure profiles change little under the influence of external sources of heat and particles. The plasma density profile, n, can change more freely.
c. In steady state the profiles of temperature, pressure and toroidal rotation velocity are close to the canonical profiles.

d. Pure radial heat pinch, if present, is very small. It is realized although the plasma is an open system with respect to energy and thus in principle the second law of thermodynamics could not be fulfilled.
e. There is a radial particle pinch and a pinch of the toroidal rotation momentum.

To construct the expression for the fluxes of the various quantities, it is convenient to use the notion of critical gradient that is defined as the logarithmic derivative of the canonical profiles. For the quantitative estimation of the gradients of the canonical profiles, we will use the dimensionless quantities

$$\frac{R_0}{L_{Tc}} = -R_0 \frac{d}{d\rho}(\ln T_c) = -R_0 \frac{T_c'}{T_c},$$
$$\frac{R_0}{L_{pc}} = -R_0 \frac{d}{d\rho}(\ln p_c) = -R_0 \frac{p_c'}{p_c},$$
$$\frac{R_0}{L_{\omega c}} = -R_0 \frac{d}{d\rho}(\ln \omega_c) = -R_0 \frac{\omega_c'}{\omega_c}. \tag{5.7}$$

It follows from conditions (a)–(e) that the fluxes should be defined such that the leading term determines the relaxation of the profiles of temperature, pressure, and angular frequency towards the canonical profiles. However, the definition of the heat flux and the definition of the particle flux must be different, because of the difference in the experimental properties (d) and (e).

Note the following asymmetry in the behavior of temperature profile. The property (d) means that if the temperature gradient exceeds the critical gradient, the effective heat diffusivity should increase significantly, preventing further deviations of the temperature profile from the canonical one. In the opposite case, where the temperature gradient is less than the critical gradient (in the case of small heat fluxes), there is no relaxation of the flat temperature profile to a peaked canonical one due to the property (d). This situation occurs typically with off-axis heating of the plasma as in that case in the central part of the plasma both the temperature gradients and heat fluxes are small. On the contrary, in the region outside of the power deposition area, the temperature gradients are above the critical gradients, and the heat fluxes and the effective heat diffusivities are high.

For the particle flux and angular momentum flux, the definition of the fluxes has to be symmetric with respect to the canonical profiles. This means that when crossing the critical gradient the flux changes sign. Therefore, the relaxation to the canonical profile occurs for both more flat and more peaked profiles of pressure and momentum.

The requirement of a linear dependence of the fluxes on the gradients of temperature, pressure and angular velocity determines the heat fluxes uniquely. The heat fluxes for the electron ($k=e$) or ion ($k=i$) channels are defined as follows [2–5]:

$$q_k = -\kappa_k^{PC} T_k \left(\frac{T_k'}{T_k} - \frac{T_c'}{T_c} \right) H \left[-\left(\frac{T_k'}{T_k} - \frac{T_c'}{T_c} \right) \right] - \kappa_k^0 T_k' + \frac{3}{2} T_k \Gamma_n \tag{5.8}$$

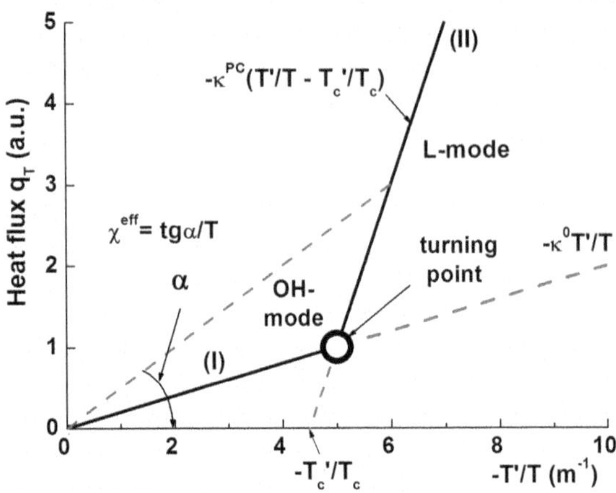

Fig. 5.1 The dependence of the heat flux q_T on the relative temperature gradient T'/T at a fixed radial point

Here $T' = \frac{\partial T}{\partial \rho}$, $\kappa_k^{PC} = n\chi_k^{PC}$ is the stiffness of the temperature profile, the index *PC* means "Profile Consistency", $H(x)$ is the Heaviside function, $H(x)=1$, if $x \geq 1$, $H(x)=0$ for $x<1$. The symbol κ_k^0 represents the thermal conductivity, determined by processes not related to the effect of profile consistency (for example, neoclassical effects, or MHD mixing during sawtooth oscillations, etc.). The latter (convection) term in the flux (5.8) is determined by the particle flux Γ_n (see below).

Because of the Heaviside function in the first term of (5.8) this term vanishes when $\left|\frac{T_k'}{T_k}\right| < \left|\frac{T_c'}{T_c}\right|$, which is the expression of the fact that heat cannot propagate against the temperature gradient. The coefficients κ_k^{PC} and κ_k^0 have to be determined from the experimental data. It should also be borne in mind that $\kappa_k^{PC} \gg \kappa_k^0$ due to the experimental property (c).

For plasmas with on-axis heating, the main term in the expression (5.8) for the heat flux is the first term containing the product $\kappa_k^{PC} T_k \left(\frac{T_k'}{T_k} - \frac{T_c'}{T_c}\right)$. It determines the conservation of the shape of the temperature profile (the "stiffness" of the profile) when changing the total deposited heating power, plasma density and edge temperature. If the profile of the deposited heating power changes and becomes peaked at the edge, the temperature profile may lose its stiffness due to the presence of the Heaviside function in the heat flux.

To simplify the quantitative characteristics of heat transport it is common to use the so-called "effective" heat diffusivity

$$\chi_k^{eff} = -\frac{q_k}{n\frac{\partial T_k}{\partial \rho}} \tag{5.9}$$

The value of χ_k^{eff} is usually several times lower than $\chi_{e,i}^{PC}$ since the terms within the first brackets in (5.8) partially cancel out. In the discharges with low deposited power (e.g., in Ohmic discharges) the mutual compensation reaches 90–95 %.

Figure 5.1 illustrates the formulas (5.8) and (5.9). Here the dependence of the heat flux (5.8) on the relative temperature gradient T'/T is shown at a fixed point in space. To simplify the figure we have omitted the index k and the convective term $\frac{3}{2}T_k\Gamma_n$. The dependence on the relative temperature gradient T'/T consists of two branches. For small temperature gradients, i.e. when $\left|\frac{T_k'}{T_k}\right| < \left|\frac{T_c'}{T_c}\right|$ the effective heat conductivity is low, because it is determined by the second term of (5.8). This part of the dependence $q_T(T'/T)$ will be called the first branch (I). If the temperature gradient is sufficiently large, i.e. $\left|\frac{T_k'}{T_k}\right| > \left|\frac{T_c'}{T_c}\right|$, then the heat flux is determined by the steep second branch (II). The L-mode experiments follow this branch.

Ohmic discharges are located in this diagram in the vicinity of the turning point in Fig. 5.1. At low plasma density in Ohmic discharges, the electron temperature and, therefore, the plasma conductivity are relatively high. Therefore the deposited power is low and the process follows the first branch (I). When the plasma density is increased, the electron temperature decreases and the operating point in Fig. 5.1 may shift to branch (II) via the turning point. In this case, the heat diffusivity rapidly increases. Apparently, this property is linked to the saturation of the energy confinement time in Ohmic discharges at a sufficiently high density.

Since the pressure profile in the experiment is well-preserved, while the density profile can vary greatly depending on the experimental conditions, we adopt for the particle flux the following expression [6]

$$\Gamma_n = -D^{PC} n\left(\frac{p'}{p} - \frac{p_c{}'}{p_c}\right) - D^0 n' + \Gamma_n^{neo}, \quad p = n(T_e + T_i). \qquad (5.10)$$

Here Γ_n^{neo} is a neoclassical particle flux and note that $D^{PC} \gg D^0$. As $\frac{p'}{p} = \frac{n'}{n} + \frac{(T_e+T_i)'}{T_e+T_i}$ the expression for the particle flux (5.10) includes, in addition to particle diffusion, also a term for thermal diffusion. The term containing $p_c{}'/p_c$, corresponds to the convection (pinching) of the particles. If the temperature profiles of the electrons and ions are similar $\left(\frac{T_e'}{T_e} = \frac{T_i'}{T_i} = \frac{T'}{T}\right)$ then $\frac{(T_e+T_i)'}{T_e+T_i} = \frac{T'}{T}$ independent of the ratio T_e/T_i. Note that in the expression for the particle flux (5.10) the Heaviside function is absent. This means that if the condition $\left|\frac{p'}{p}\right| < \left|\frac{p_c{}'}{p_c}\right|$ is satisfied, the particle pinch is greater than the particle diffusion. The second term in (5.10) corresponds to the "background" particle flux. It takes into account a possible turbulent flux, independent of the canonical profiles.

Figure 5.2 shows the dependence of the particle flux on the pressure gradient. For simplicity, we omit here the second and third terms in the expression for the flux (5.10). In contrast to heat conduction, where the energy is mainly deposited in the central part of the plasma, the main source of particles is the influx of cold neutrals from the first wall into the plasma. If the plasma density is sufficiently high, the capture of particles through ionization and charge exchange occurs in the edge layers of the plasma and the particle source in the central part of plasma under steady state conditions is close to zero. This is illustrated in Fig. 5.2 by the words "plasma core" and "plasma edge" which mark the radial position of the corresponding points. Particle fluxes in these points are quite different. During the non-stationary stage of the

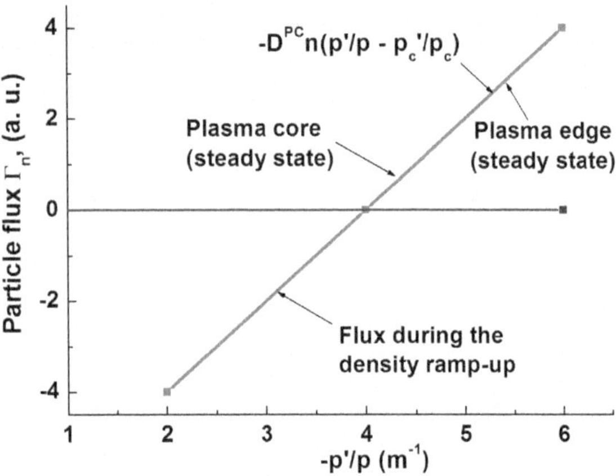

Fig. 5.2 The dependence of the main part of the particle flux Γ_n on the relative pressure gradient p'/p

density ramp up the particle flux can become negative and this is also marked in Fig. 5.2. In the case of plasma heating by neutral beam injection, which is a source of both energy and particles, the discharge operating point in Fig. 5.2 is shifted to the region of positive particle fluxes.

Finally, for the flux of toroidal rotation momentum, we will assume the following expression [7]

$$q_L = q_\omega = nm_i R_0^2 \chi_\omega^{PC} \omega \left(\frac{\omega'}{\omega} - \frac{\omega_c{}'}{\omega_c} \right) - nm_i R_0^2 \chi_\omega^0 \omega'. \tag{5.11}$$

To simplify the expression for the momentum flux the density n is taken away from the radial derivative (only $n\omega'$ is included into (5.11), not the product $(n\omega)'$). This is partly justified by the fact that we have a separate Eq. (5.1) for density and the expression (5.10) for the particle flux. It is also implicitly assumed that $\chi_\omega^{PC} \gg \chi_\omega^0$. As in the expression for the particle flux, the Heaviside function in (5.11) is missing, and thus the momentum pinch can exceed the momentum diffusion.

Note the paradoxical property of the system (5.1–5.4) and fluxes (5.8) (5.10), (5.11). Canonical profiles were found in the previous chapters for the characteristics of poloidal magnetic field μ_c and j_c. At the same time, the transport Eq. (5.3) for the poloidal flux ψ does not contain the canonical profiles, and the current conductivity $\sigma_{||}$ is classical or neoclassical. On the contrary, in the heat and particle fluxes (5.8–5.11) the main parts are the "turbulent" fluxes, containing the critical gradients formed from the canonical profiles. Thus, during the evolution of the temperature and density the profiles of μ and j relax to the canonical profiles μ_c and j_c through the influence of "normal" conductivity $\sigma_{||}$, depending on the electron temperature profile, which is determined by the canonical profile T_c.

But the skin time for current diffusion is quite long and very often the current profile is far from the canonical current profile during the whole of the discharge.

5.3 Definition of the Transport Coefficients

The transport coefficients $\kappa_k^{PC} = n\chi_k^{PC}$, $\kappa_k^0 = n\chi_k^0$ $(k{=}e, i)$, D^{PC}, D^0, χ_ω^{PC} and χ_ω^0, determining the rate of relaxation, have to be defined by comparison the calculations with experiment. Typically, the ratio $\kappa_k^{PC}/\kappa_k^0 = \chi_k^{PC}/\chi_k^0$ is in the range $\chi_k^{PC}/\chi_k^0 = 6-15$. This means that the set of transport Eqs. (5.1–5.4) is stiff due to the presence of a large parameter in Eq. (5.2). The diffusion coefficient D^{PC} is usually several times smaller than the coefficient χ_k^{PC}; thus the pressure profile is not as stiff as the temperature profile when the temperature gradient is greater than the critical one. The coefficients χ_k^0 and D^0 will be defined in the next chapter, when we will discuss the non-linear version of the CPTM. For the main coefficient κ_k^{PC} we use the expression that was defined in [8] and further refined in [9]

$$\kappa_k^{PC} = \left\{\frac{\alpha_k}{MA^{3/4}} q(\rho = \frac{\rho_{max}}{2}) q_{cyl}(\rho = \rho_{max}) T_k^{1/2}(\rho = \frac{\rho_{max}}{4})\frac{\bar{n}}{B}\right\}\left\{\left(\frac{3}{R_0}\right)^{1/4}\right\} \tag{5.12}$$

It is assumed here that κ_k^{PC} is independent of ρ and the following "practical" units are used: κ_k^{PC} in 10^{19} m^{-1}s^{-1}, $\alpha_e=3.5$, $\alpha_i=5$, M is the relative ion mass, T_k is the temperature in keV, $\bar{n}$ is the line averaged plasma density in 10^{19} m^{-3}, B is the toroidal magnetic field in T, R_0 is the major plasma radius in m, $q_{cyl}=5a^2B/(R_0I)$. For devices with a moderate aspect ratio $A=R_0/a=3$–5 the value of κ_k^{PC} is in the range $\kappa_k^{PC} \sim 5-15$ (10^{19} m^{-1}s^{-1}). For devices with low A and low magnetic field B (mainly spherical tokamaks) $\kappa_k^{PC} \sim 20-30$ (10^{19} m^{-1}s^{-1}).

The expression for κ_k^{PC} in (5.12) is divided into two factors in curly brackets. The first factor is obtained from the analysis of experiments on several tokamaks (JET, TFTR, ASDEX, T-10) with a sufficiently high aspect ratio [8]. The second factor [9] is given by the limitations (4.89). In fact, in expression (5.12), the first factor is given by

$$(\chi_k^{PC})_1 \sim \frac{q^2}{A^{3/4}}\frac{T^{1/2}}{B}. \tag{5.13}$$

In the approximation of a quasi-neutral plasma the parameters q and A, by virtue of (4.60) are invariant under the transformation (4.47). Using (4.89) the expression (5.13) can also be written as:

$$(\chi_k^{PC})_1 \sim \frac{q^2}{A^{3/4}}\frac{T^{1/2}}{B} \sim a^2BF, \tag{5.14}$$

where F is a function of the invariants (4.58). We thus find that the ratio $T^{1/2}/(aB)^2$ would also need to be a function of the invariants (4.58). Unfortunately, the first factor of (5.12), found in [8] from the comparison of the calculations with experiment, does not satisfy this condition and must be corrected. The second factor in (5.12) represents this correction, because the ratio

$$\frac{T^{1/2}}{a^2 B^2 R_0^{1/4}} = \frac{J_2^{1/4}}{J_3^{1/2} A^{1/4}} \tag{5.15}$$

is a function of the invariants (4.58). Of course, the correction factor selected here is not unique, but it is convenient for our purposes, since for the each separate device the parameter R_0 is almost fixed, and the factor $R_0^{1/4}$ changes minimally for a large range in R_0 values.

In the model of particle transport, we assume that the time behavior of the line-averaged density is given. Thus, the source of the particles (in the form of an influx of cold neutrals from the wall q_N) is determined such that the calculated line averaged density corresponds to the given value. As transport of both energy and particles is determined by the same type of turbulence, the dependence of the associated transport coefficients on the plasma parameters should also be similar. Thus for the diffusion coefficient D^{PC} (determining the stiffness of the pressure profile) we assume the following relation [6]

$$D^{PC} = C_n \chi_e^{PC} \tag{5.16}$$

The value of the coefficient C_n is determined by the magnitude of the flux of cold neutrals q_N. This value is usually not measured, so we are using here a rough estimate of this flux. This estimate is based on the usual assumption that q_N increases with density and with deposited power; it has a value in the range $q_N = 2\times 10^{21}$–5×10^{22} s^{-1} for the devices of scale T-10–JET. We have to choose the coefficient C_n so that the flux of neutral particles is in the above-mentioned range. From the analysis of experiments on several tokamaks [6] we find that the value of the coefficient C_n is in the range

$$C_n = 0.08 - 0.1 \tag{5.17}$$

We assume that the stiffness of the profile of the angular rotation frequency χ_ω^{PC} is proportional to the stiffness of the electron temperature profile:

$$\chi_\omega^{PC} = C_\omega \chi_e^{PC}, \tag{5.18}$$

where χ_e^{PC} is defined by (5.12). The analysis of experiments on JET has shown that [7]

$$C_\omega = 0.5/\bar{n}^{1/3}, \tag{5.19}$$

where $\bar{n}$ is the line-averaged plasma density. This shows in particular that in high density plasmas the stiffness of the rotation profile is noticeably less than the stiffness of the temperature profile.

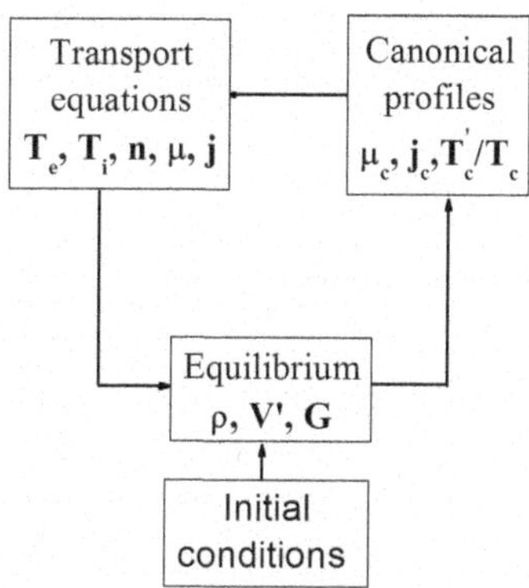

Fig. 5.3 The structure of the transport model

5.4 The Structure of the Full Transport Model

The structure of the full transport model is shown in Fig. 5.3. The model contains an initialization module where the initial profiles of the plasma density, electron and ion temperatures, frequency of rotation and the toroidal current density are set. It is followed by three main calculation modules. In the first module the Grad–Shafranov equation is solved to determine the plasma equilibrium, the geometry of the magnetic surfaces together with the coordinate mapping and the necessary metric coefficients V', G and G_1. In the second module the canonical profiles μ_c, j_c and critical gradients are obtained by solving the canonical profile Eq. (2.98) with boundary conditions (2.99–2.101). Finally, the third module solves the set of transport equations, which determine the electron and ion temperatures T_e and T_i, the plasma density n, the angular rotation frequency ω, the function μ and current density j.

Note that in this model the canonical profiles can also change in time. This is due to changes in the current profile, in the plasma equilibrium and in the metric coefficients. However, the relaxation rate of the temperature and plasma density profiles is much higher than the rate of change of current density and canonical profiles, and therefore, sooner or later, the real profiles of plasma parameters converge to the canonical profiles. The presence of sources of heat, particles and momentum do not allow the computed profiles to coincide completely with the canonical ones. In discharges with large values of the additional heating power the difference between the experimental temperature profiles and the canonical profiles can be significant. This will be discussed in the next chapter.

5.5 Stiffness of the Heat Diffusion Equation

Let us estimate for which values of the plasma parameters the Eq. (5.2) can be considered as stiff. We assume steady state conditions for simplicity. Integrating (5.2) over the plasma volume, dropping the index k, and leaving only the main first term we obtain

$$q_T = \frac{1}{G_1 V'} \int_0^{\rho} V' P d\rho = -\kappa^{PC} T \left(\frac{T'}{T} - \frac{T_c'}{T_c} \right) = -\kappa^{PC} T' \Delta \quad (5.20)$$

We define

$$\Delta = 1 - \left(\frac{T_c'}{T_c} \right) \left(\frac{T'}{T} \right)^{-1}, \quad Q(\rho) = \frac{1}{G_1 V'} \int_0^{\rho} V' P d\rho. \quad (5.21)$$

Here $Q(\rho)$ is the power deposited inside the radius ρ. If the temperature profile $T(\rho)$ is sufficiently close to the canonical profile $T_c(\rho)$, then

$$\Delta \ll 1. \quad (5.22)$$

Condition (5.22) is called the condition of stiffness of the Eq. (5.2). In view of (5.20–5.21), the stiffness condition (5.22) can also be written:

$$\frac{Q}{\kappa^{PC} |T'|} \ll 1. \quad (5.23)$$

By (5.22),

$$\frac{T'}{T} \approx \frac{T_c'}{T_c}, \quad (5.24)$$

therefore, instead of (5.23), we have:

$$\frac{Q}{\kappa^{PC} T |T_c'/T_c|} \ll 1. \quad (5.25)$$

Since $Q(\rho)$ is increases with increasing radius, the largest value for (5.25) is obtained at the plasma boundary

$$\frac{Q_a}{\kappa^{PC} T |T_c'/T_c|} \ll 1, \quad Q_a = Q(\rho_{max}). \quad (5.26)$$

Since $|T_c'/T_c| \sim 1/a$, we get for monotonous temperature profiles

$$\kappa^{PC} \gg \frac{a Q_a}{T_0}, \quad (5.27)$$

where $T_0 = T(0)$. We now recognize that $Q_a = Q_{tot}/S$, where Q_{tot} is the total deposited power, and S is the total plasma surface. Thus we finally obtain for the stiffness condition

$$\kappa^{PC} \gg \frac{a Q_{tot}}{S T_0}. \tag{5.28}$$

In practical units, the inequality (5.28) becomes:

$$\kappa^{PC}[10^{19}\text{m}^{-1}\text{s}^{-1}] \gg 625\frac{a[\text{m}]Q_{tot}[\text{MW}]}{S[\text{m}^2]T_0[\text{keV}]}. \tag{5.29}$$

Using the approximate expression $S \approx 4\pi^2\, kaR_0$, we obtain:

$$\kappa^{PC}[10^{19}\text{m}^{-1}\text{s}^{-1}] \gg 15.6\frac{Q_{tot}[\text{MW}]}{kR_0T_0[\text{keV}]}. \tag{5.30}$$

For T-10 (Ohmic plasmas Q_{tot} = 0.5 MW, $T(0)$ = 1 keV, R_0 = 1.5 m, k = 1, q ~ 3) and JET (NBI heated plasmas, Q_{tot} = 10 MW, $T(0)$ = 10 keV, R_0 = 3 m, k = 1.6, q_{cyl} ~ 2), the condition (5.30) becomes:

$$\kappa^{PC} \gg 5 \times 10^{19}\text{m}^{-1}\text{s}^{-1}(\text{for T-10}), \quad \kappa^{PC} \gg 3 \times 10^{19}\text{m}^{-1}\text{s}^{-1}(\text{for JET}) \tag{5.31}$$

Estimating the value of κ^{PC} using (5.12), we find: κ^{PC}(T-10) ~ 5, κ^{PC}(JET) ~ 15. Thus, for typical JET NBI heated discharges, Eq. (5.2) is stiff; for T-10 Ohmic discharges the stiffness of Eq. (5.2) is small (at low values of q ~ 3).

5.6 Examples

The following should be taken into account to assess properly the quality of modeling:

1. When using a database in which the experimental data are processed by the TRANSP code (PPL, Princeton University) the data are usually smoothed, and thus there is no direct information on the experimental errors in the data.
2. The profiles of the heating power deposition from NBI given in databases are the result of calculations. This is in particular true for the ITER database [10].

In most of the cases, the Monte Carlo code NUBEAM is used. This code takes into account the plasma, beam and tokamak geometry, the details of the energy and flux of the beam neutrals, the orbit and charge-exchange losses etc. As a result the errors in the calculations of the power deposition profile are difficult to estimate. In discharges with tangential injection, sufficiently high toroidal magnetic field and moderate plasma densities, usually about 80 % of the injected power is absorbed by plasma. In low aspect ratio plasmas, at low magnetic field and high density the absorbed neutral beam power reduces to less than 50 %.

In what follows, we illustrate applications of the transport model (5.1–5.4) with fluxes as defined in (5.8–5.11).We did not integrate Eq. (5.1) for the density when

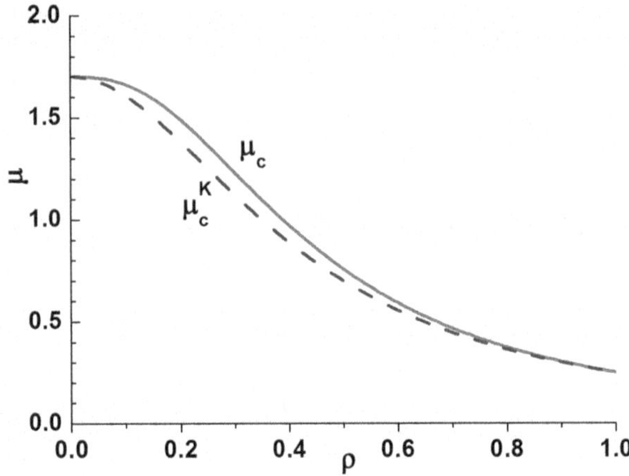

Fig. 5.4 The canonical profile $\mu_c(\rho)$, found by general model (2.98–2.101), and the Kadomtsev canonical profile $\mu_c^K(\rho)$ (2.40) for the one of Ohmic discharges of T-10

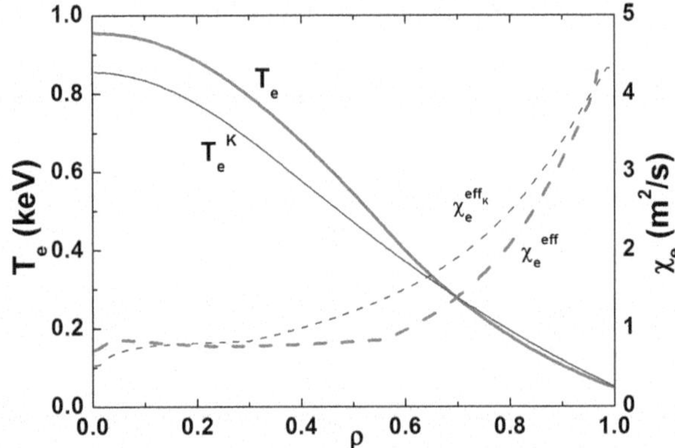

Fig. 5.5 The electron temperature profiles $T_e(\rho)$ and the profiles of the effective heat diffusivities χ_e^{eff}, calculated by the model (5.1–5.4), (5.8–5.10), for both canonical profiles shown in Fig. 5.4

the experimental density profile was available. Discharge data were taken from the JET, MAST and T-10 databases.

We start with a comparison of canonical profiles μ_c, calculated by the general formulas (2.98–2.101) of Chap. 2, and the approximate canonical profile μ_c^K (2.40), proposed by Kadomtsev. Figure 5.4 shows these canonical profiles for one of the discharges of T-10 which has a circular plasma cross-section. Note that the profile of μ_c is only slightly broader than the profile proposed by Kadomtsev. The impact of this difference on the energy balance is shown in Fig. 5.5, showing the profiles of the electron temperature obtained using both canonical profiles. It is clear that a small broadening of the canonical profile leads to a noticeable increase in the central electron temperature (in this case more than 100 eV). We also show the profiles of the effective electron heat diffusivity. Most of the differences between the two profiles is located mainly in the gradient zone $0.5<\rho/\rho_{max}<0.8$.

Figure 5.6a and b shows the experimental and calculated profiles of the electron and ion temperatures, $T_{e,i}^{exp}$ resp. $T_{e,i}$, for DIII-D discharge #71384. This discharge

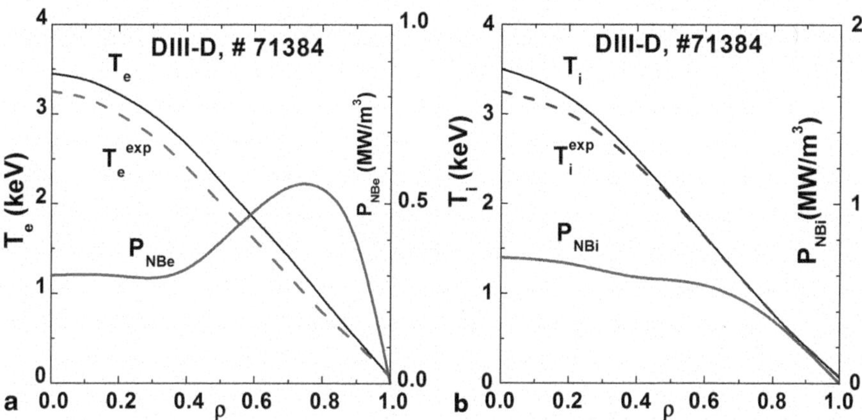

Fig. 5.6 The experimental and calculated profiles of the electron temperature, T_e^{exp} and T_e, (**a**) and ion temperature, T_i^{exp} and T_i, (**b**) for the DIII-D discharge #71384 with high density $\bar{n} = 9.6 \times 10^{19}\,\mathrm{m}^{-3}$ and high power of the NBI heating P_{NB} = 14 MW. The profiles of calculated power absorbed by the electrons and ions, P_{NBe} and P_{NBi}, are also shown

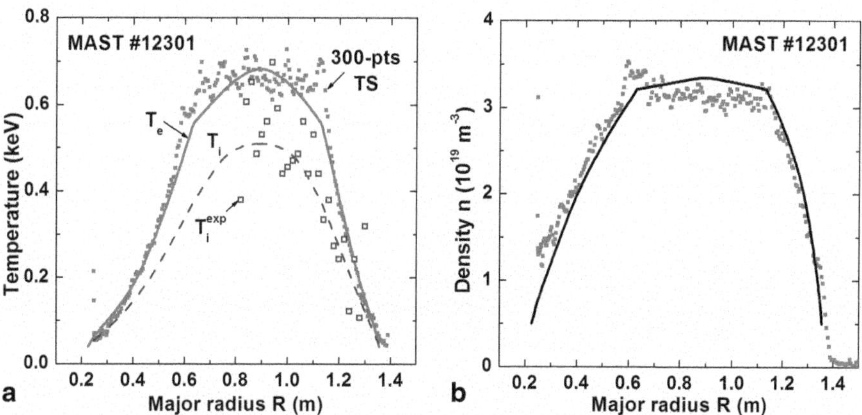

Fig. 5.7 The experimental and calculated temperature profiles of electrons and ions, T_e and T_i, (**a**) and profiles of the plasma density n (**b**) for the MAST Ohmic discharge #12301 with moderate density. The measurements of T_e and n are provided by Thomson scattering diagnostics (full *red* squares. The measurements of the ion temperature T_i by CXRS diagnostics are marked by open (blue) squares. Calculations: the *solid curves* for T_e and n, and *dotted lines* for T_i

is characterized by a high density $\bar{n} = 9.6 \times 10^{19}\,\mathrm{m}^{-3}$, strong NBI injection with power $P_{NB} \sim 14$ MW, a moderate value of the parameter $q_a = 3.7$, plasma elongation $k = 1.6$, and triangularity $\delta = 0.16$. This figure shows also the profiles of power deposition to electrons and ions. Because of the high plasma density the neutral beam deposition to electrons takes place mainly in the plasma edge.

Figures 5.7 and 5.8 show the simulation results for Ohmic discharges of MAST. Figure 5.7 shows the experimental and calculated profiles of electron and ion tem-

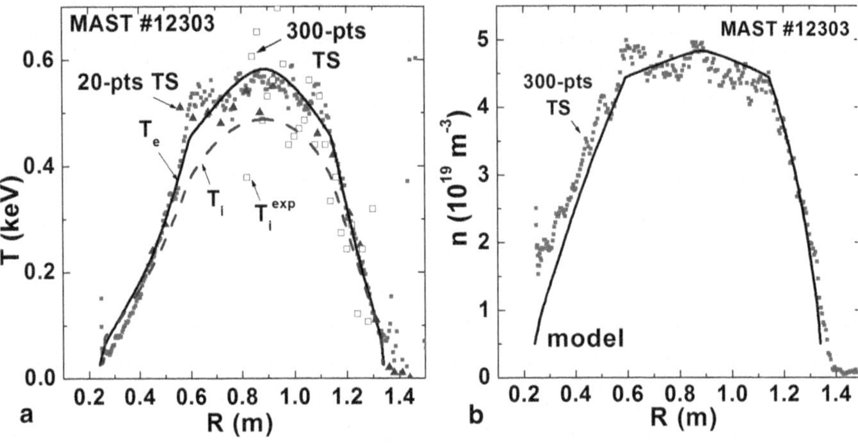

Fig. 5.8 The same as Fig. 5.7, but for the MAST discharge #12303 with a high density

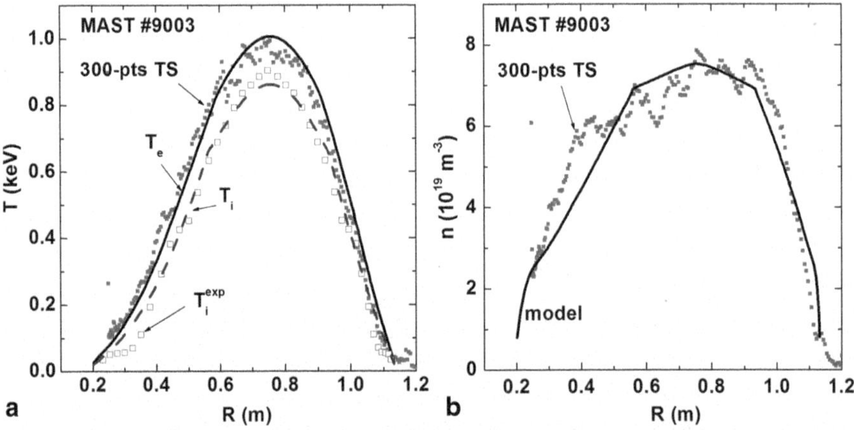

Fig. 5.9 The same as Fig. 5.7, but for the MAST discharge #9003 with NBI heating

peratures (a) and density (b) for discharge #12301 with moderate density. The measurements of the electron temperature and plasma density are performed by Thomson scattering (TS) diagnostics. The ion temperature is measured by the CXRS diagnostic. This diagnostic measures the ion temperature only along the outer half of the minor radius. Figure 5.8 shows modeling results for MAST discharge #12303, which had a markedly higher density. In this case the difference between the temperature profiles for electrons and ions is considerably less.

Figure 5.9a, b shows the simulation results for MAST discharge #9003 with NBI heating. There are no primary experimental data for the ion temperature, but the data were processed by the code TRANSP. This code maps also the experimental CXRS diagnostic data from the low field side to the high field side of the plasma,

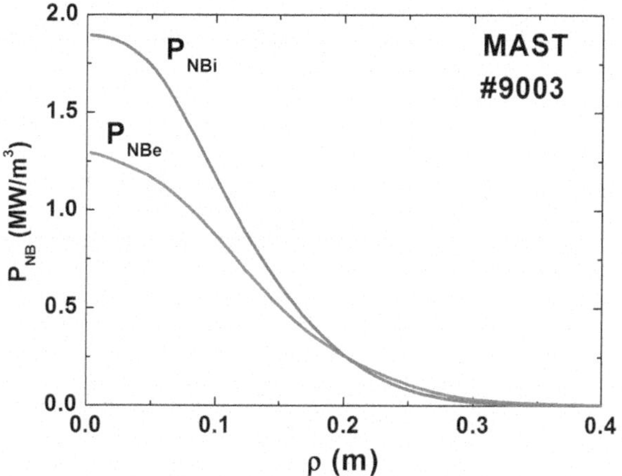

Fig. 5.10 The profiles of NBI power P_{NB}, deposited to electrons P_{NBe} and to ions P_{NBi}, in the MAST discharge #9003. Profiles are calculated by code TRANSP

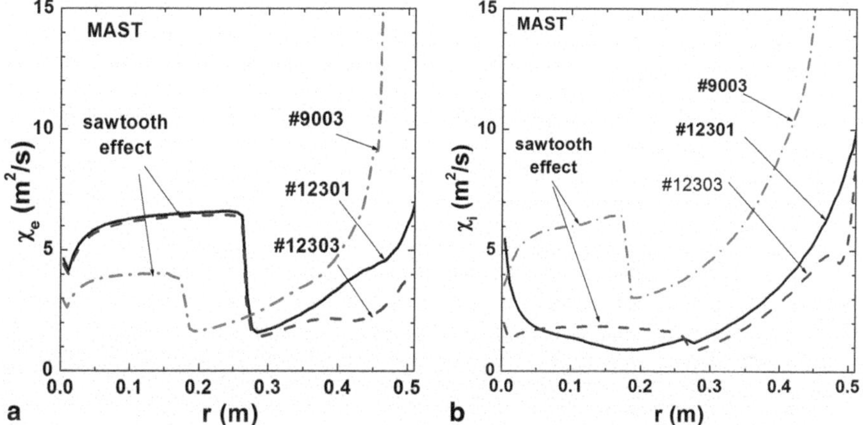

Fig. 5.11 The calculated profiles of effective heat diffusivity for electrons (**a**) and for ions (**b**) for MAST discharges, shown in Figs. 5.7–5.9. Plateau in the central part of curves shows the magnitude of the effective heat diffusivity χ^{MHD}, used for the averaging description of the effect of sawtooth oscillations inside the region $r<r_s$, where $q(r_s)\approx 1$

as shown in Fig. 5.9a. Figure 5.10 shows the power deposition profiles to electrons and ions due to NBI as calculated by TRANSP. The simulation results plotted in Figs. 5.7–5.9 show that standard deviations between the experimental and calculated data do not correctly assess the accuracy of the simulation because they include also the experimental errors.

Figure 5.11 illustrates the calculated profiles of the effective heat diffusivities for the electrons and ions for MAST discharges, shown in Figs. 5.7–5.9. The fact that the curves are flat in the central part of the plasma corresponds to an artificial increase of κ_k^0 needed to describe the average effect of sawtooth oscillations in the region $\rho<\rho_S$, where $q(\rho_S)\approx 1$. The plots show clearly that NBI heating of the

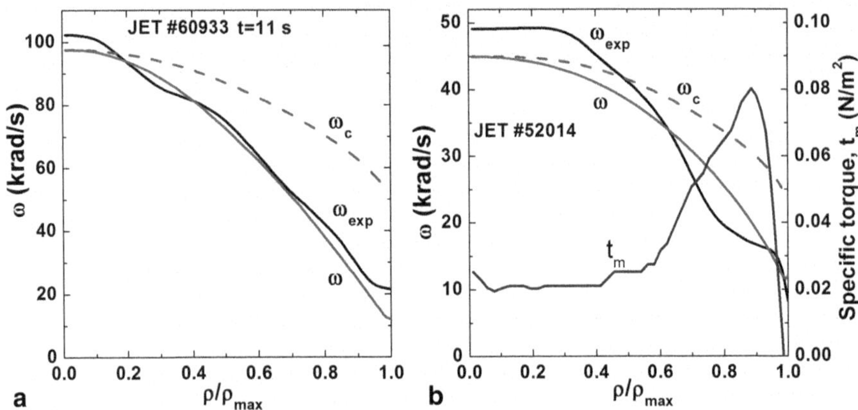

Fig. 5.12 The experimental and calculated profiles of the toroidal rotation angular frequency ω for JET discharges #60933 and #52014, and the canonical profiles ω_c. For a discharge #52014 it is shown also the profile of deposited torque t_m, generated by the longitudinal component of the hot neutrals beam. Due to the high plasma density the torque profile is skinned, however, the pinch of the moment allows to save a monotonically decreasing profile of the rotation velocity

plasma leads to an increase in the effective heat diffusivity over the whole plasma cross section.

The next set of examples show modeling results related to plasma rotation. Figure 5.12 shows the experimental and calculated profiles of the angular frequency of the toroidal rotation $\omega(\rho)$ for JET discharges #60933 and #52014. The canonical profiles of the angular frequency ω_c are also plotted. For a discharge #52014 also the profile of the deposited torque t_m, generated by the longitudinal component of the hot neutrals beam is shown. In this discharge the plasma density is high ($\bar{n} \sim 10.5 \times 10^{19}\,\mathrm{m}^{-3}$), therefore the profile of t_m is peaked at the edge, just like the particle and energy deposition profile of NBI. However, the pinch of the rotation moment, included in the expression for the flux (5.11), allows us to reproduce the monotonically decreasing rotation velocity profile as a function of plasma radius.

Figure 5.13 shows the same profiles as in Fig. 5.12, but for MAST discharge #13035. Note that the absolute values of the angular frequency in MAST are much higher than in JET, although the beam power at MAST is significantly lower than at JET. This is because the major plasma radius in MAST, $R_0 = 0.8$ m, is much less than the major plasma radius at JET: $R_0 = 3$ m, and the moment of inertia of plasma in MAST is much smaller than the moment of inertia of JET plasmas.

As a next example we show simulation results for a non steady-state discharge. Figure 5.14 shows the time evolution of the plasma current $I_p(t)$ and line-averaged density $\bar{n}$ for MAST Ohmic discharge #24433. The goal of the simulation was to clarify the character of the longitudinal conductivity of the plasma, i.e. whether it was sufficiently close to the neoclassical conductivity. We assumed in these calculations that the conductivity was purely neoclassical. Two models were used in calculations. First we used the so-called MEXP model, which is also the simplest one: the experimental electron and ion temperature profiles were used, and only Eq. (5.3)

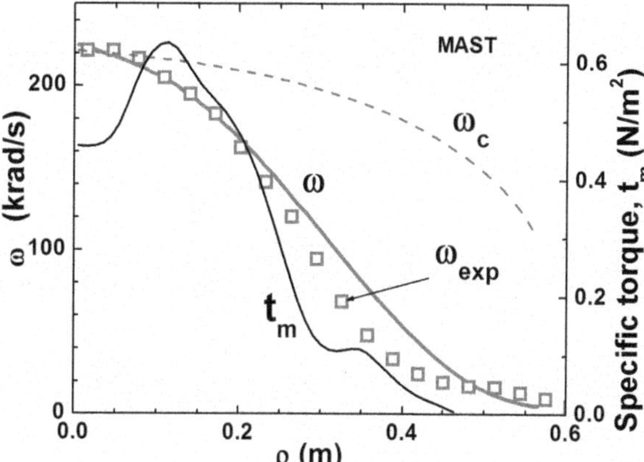

Fig. 5.13 The same profiles as in Fig. 5.12, but for the MAST discharge #13035. Also shown are the profiles of the deposited torque t_m and of the canonical toroidal rotation angular frequency ω_c

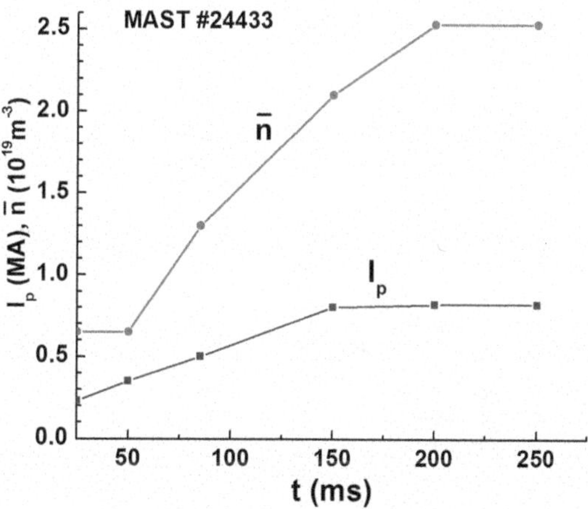

Fig. 5.14 The scenarios of plasma current I_p and chord averaged density $\bar{n}$ for the MAST discharge #24433

for the potential ψ was integrated. As a second model the full linear version of the CPTM (5.1–5.3) was used.

Figure 5.15 shows the time evolution of the experimental and computed central electron temperature. The time evolution of q in the plasma centre $q(0, t)$, found experimentally and calculated by the two models MEXP and CPTM, is shown in Fig. 5.16. Figure 5.17a, b shows the experimental and calculated profiles of the parameter q at the beginning (a) and at the end (b) of the discharge ($t=0.03$ s in (a) and $t=0.25$ s in (b)). It is evident that during the discharge the value of parameter q in the plasma core changes over more than a factor of 10. Notwithstanding this, the calculated profiles at all times correspond quite well to the experiment. This means that the plasma conductivity is very close to the neoclassical one.

These examples show that the proposed model adequately describes a wide range of the Ohmic discharges and discharges with NBI heating, allowing a detailed analysis of the heat and particle fluxes and transport coefficients.

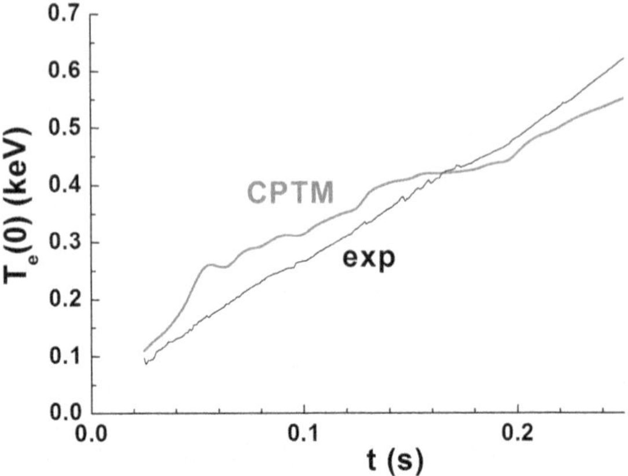

Fig. 5.15 The time behavior of the experimental and calculated central temperature of the electrons in the MAST discharge # 24433

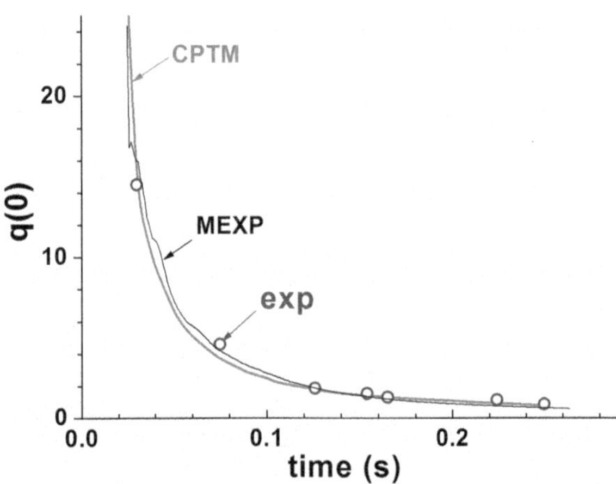

Fig. 5.16 The evolution of safety factor q in the plasma centre measured in the experiment and calculated according to two models: the CPTM and the MEXP, the latter is the simplified model, in which the temperature profiles of electrons and ions are taken from the experiment, and only the Eq. (5.3) is integrated. In both models, the conductivity is assumed to be the neoclassical one

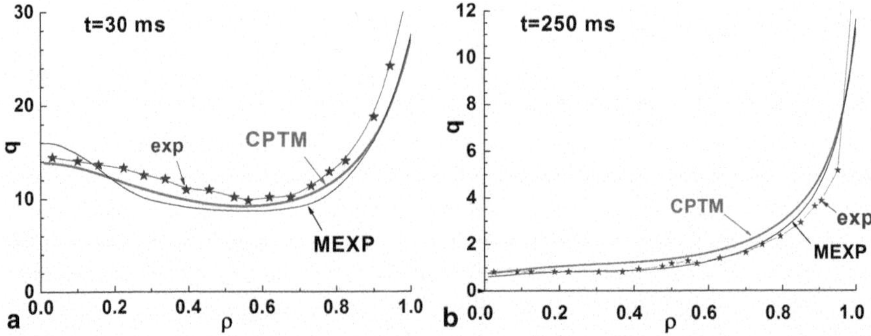

Fig. 5.17 The experimental and calculated profiles of safety factor $q(\rho)$ at the time moment t=0.03 s (**a**) and t=0.25 s (**b**)

References

1. Pereverzev, G.V., Yushmanov, P.N.: ASTRA: Automated System for Transport Analysis in a Tokamak. Max-Planck IPP Report. IPP 5/98. Garching, Germany, February 2002
2. Dnestrovskij, Y.N., Pereverzev, G.V.: Energy confinement in the T-10 Tokamak and canonic profile model. Plasma Phys. Control Fusion **30**, 1417 (1988)
3. Dnestrovskij, Y.N., Dnestrovskij A.Y., et al.: Application of the canonical profile theory to the problems of heat transport in Tokamaks. Plasma Phys. Rep. **30**, 1 (2004)
4. Dnestrovskij, Y.N., Dnestrovskij, A.Y., et al.: Stiffness and rigidity of the temperature profile in a Tokamak. Plasma Phys. Rep. **30**, 717 (2004)
5. Dnestrovskij, Y.N., Dnestrovskij, A.Y., et al.: Self-organization of plasma in Tokamaks. Plasma Phys. Rep. **31**, 529 (2005)
6. Dnestrovskij, Y.N., Razumova, K.A., et al.: Self-consistency of pressure profiles in Tokamaks. Nucl. Fusion **46**, 953 (2006)
7. Dnestrovskij, Y.N., Andreev, V.F., et al.: Canonical profiles and transport model for the toroidal rotation in Tokamaks. Plasma Phys. Control Fusion **53**, 085025 (2011)
8. Dnestrovskij, Y.N., Berezovskij, E.L., et al.: Transport model of canonical profiles for electron and ion temperatures in Tokamaks. Nucl. Fusion **31**, 1877 (1991)
9. Dnestrovskij, Y.N., Gryaznevich, M.P., et al.: Simulation of START shots with the canonical profile transport model. Plasma Phys. Rep. **26**, 539 (2000)
10. Roach C.M., et al.: The 2008 public release of the international multi-Tokamak confinement profile database. Nucl. Fusion **48**, 125001 (2008). http://tokamak-profiledb.ccfe.ac.uk

Chapter 6
Nonlinear Version of the Canonical Profiles Transport Model (CPTM) for Improved Confinement Regimes

Abstract A nonlinear version of the CPTM is presented in this chapter. This version can be applied to tokamak regimes with improved confinement in the presence of External and/or Internal Transport Barriers (ETB, ITB). To describe these ETB or ITB in the expressions for the fluxes as presented in Chap. 5 some multipliers are included which nonlinearly (exponentially) depend on the pressure gradients. The appearance of barriers corresponds to a bifurcation of the solution of the transport equations which is localized in space. The appearance of an ETB is described by a bifurcation parameter. The appearance of an ITB is described by a bifurcation function with radial dependence. Both the bifurcation parameter and bifurcation function are determined by comparing with experimental data. Examples are given for several tokamaks. Asymptotic expressions are derived which allow to calculate the L–H transition threshold and to estimate the pedestal temperature values. The radial dependence of the ion temperature profile stiffness and its dependence on toroidal rotation velocity are discussed.

6.1 Plasma Regimes with Improved Confinement

We make the general assumption that regimes with improved confinement are regimes with transport barriers. The regime with a transport barrier at the plasma boundary, called Edge Transport Barrier or ETB, is usually called H-mode or H-regime [1, 2]. During the transition from the L-mode to H-mode the transport coefficients decrease inside the barrier by a factor of 5–50 and outside the barrier by a factor 1.5–2. A Transport barrier can also be formed inside the plasma, and that is then called Internal Transport Barrier or ITB. Inside an ITB the transport coefficients are usually strongly reduced [3].

Internal transport barriers initially occur at the locations of the resonant surfaces $q = m/n$ with low values of m and n ($m/n = 1/1, 3/2, 2/1, 5/2, 3/1$) and at low magnetic shear $s = \rho/q\partial q/\partial\rho$. For the sake of brevity, we will call such dedicated values of m/n the *principal resonances*. To realize in practice an ITB, the discharge scenario is such that the profile of $q(\rho)$ is non-monotonic with a minimum close to one of the principal resonances. To do this, auxiliary heating is switched on during the current ramp-up. The rapidly growing electron temperature profile "freezes" the current diffusion in the plasma and the profile of $q(\rho)$ becomes non-monotonic. When the

Yu.N. Dnestrovskij, *Self-Organization of Hot Plasmas,* DOI 10.1007/978-3-319-06802-2_6,

position of the minimum value of $q(\rho)$ is close to one of the principal resonances, the transport barrier forms, and the temperature and pressure gradients rapidly increase in the vicinity of the resonant surface.

To describe the regimes with improved confinement in the CPTM, we introduce the concept of the second critical gradient. Experiments show that a second critical gradient exists for the pressure profile: if the pressure gradient in a region of the plasma exceeds a second critical gradient, a bifurcation to a new plasma state takes place and the transport barrier is formed. One could to interpret this as if the plasma "forgets" about the canonical profile inside the barrier region.

The barrier is not created at the start of the additional heating power, but only after the pressure gradient exceeds the second critical gradient. The delay between the start of the heating power and the formation of the barrier is determined by the difference between the applied heating power and a certain threshold power. The smaller this difference, the longer is the time delay before the barrier is formed.

In the H-mode, the ETB is formed simultaneously for ion and electron temperatures and plasma density. This is most likely due to the strong coupling between ions and electrons at low electron temperature.

Experimentally, the formation of an ITB can manifest itself in different ways. For example, on TFTR an ITB was observed only for the ion temperature and plasma density (i–ITB and n–ITB), but the electron transport barrier (e–ITB) was not observed. On other devices an internal barrier was also observed for the electrons and even for the toroidal rotation velocity (ω–ITB). At sufficiently high heating power an ITB can form in all transport channels. In most of the cases the spatial position of the i–ITB, e–ITB, n–ITB and ω–ITB barriers is almost identical, but it can also vary over time. In some cases two internal barriers for the ion temperature have been observed. Finally, both internal and external barriers can exist at the same time. This is then called an H-mode with internal barrier.

6.2 Heat and Particles Fluxes in Improved Confinement Regimes

The existence of ion and electron transport barriers affects the formulation of the transport model. We assume that the second critical gradient for the electron pressure is different from the one for the ion pressure. We also assume that the canonical profiles for the partial pressures of electrons and ions coincide with the canonical profile for the total pressure p_c (2.102).

We introduce the dimensionless deviation of the partial pressure p_k from the canonical pressure p_c as follows

$$z_{pk} = \frac{\rho_{\max}}{\rho} a \left(\frac{p_k'}{p_k} - \frac{p_c'}{p_c} \right) \quad (k = e, i) \tag{6.1}$$

The normalization factor ρ_{max}/ρ is introduced in (6.1) in order for the deviation z_{pk} not to vanish at the point $\rho = 0$. In line with the assumptions made earlier a transport barrier occurs in a certain region of the plasma, if the deviation (6.1) is larger than the second critical gradient z_{0k} in this region

$$\left|z_{pk}\right| > z_{0k}, \tag{6.2}$$

where $z_{0k} = z_{0k}(\rho)$ is a positive function, which has to be determined by comparing the calculations with experiment. It is clear that $z_{0k}(\rho)$ is linked to the profile of $q(\rho)$, but this link will be established below.

We now introduce the so-called *forgetting factor*

$$F_k = \exp\left(-\frac{z_{pk}^2}{2z_{0k}^2}\right) \tag{6.3}$$

and note that $F_k \approx 1$ outside the *forgetting region*, where $\left|z_{pk}\right| < z_{0k}$, and $F_k \ll 1$ within the forgetting region, where $\left|z_{pk}\right| > z_{0k}$. Within the forgetting region the first terms in the fluxes (5.8, 5.10, 5.11) should be small. This means that the electron and ion heat fluxes, the particle flux and momentum flux can now be written as follows [4–6]:

$$q_k = -\kappa_k^{PC} T_k \left(\frac{T_k'}{T_k} - \frac{T_c'}{T_c}\right) H\left[-\left(\frac{T_k'}{T_k} - \frac{T_c'}{T_c}\right)\right] F_k - \kappa_k^0 T_k' + \frac{3}{2} T_k \Gamma_n \quad (k = e, i) \tag{6.4}$$

$$\Gamma_n = -D^{PC} n \left(\frac{p'}{p} - \frac{p_c'}{p_c}\right) F_n - D^0 n', \tag{6.5}$$

$$q_L \equiv q_\omega = -nm_i R^2 \chi_\omega^{PC} \omega \left(\frac{\omega'}{\omega} - \frac{\omega_c'}{\omega_c}\right) F_i - nm_i R^2 \chi_\omega^0 \omega'. \tag{6.6}$$

Here, in accordance with experiments on TFTR [7], we suppose that

$$F_n = \exp\left(-\frac{z_p^2}{2z_{0n}^2}\right), \quad z_p = \frac{\rho_{\max}}{\rho} a \left(\frac{p'}{p} - \frac{p_c'}{p_c}\right), \quad z_{0i} < z_{0n} < z_{0e}. \tag{6.7}$$

For example

$$z_{0n} \approx \tfrac{1}{2}(z_{0e} + z_{0i}). \tag{6.8}$$

Furthermore, it is assumed that a toroidal velocity transport barrier is created during the formation of an ion temperature transport barrier.

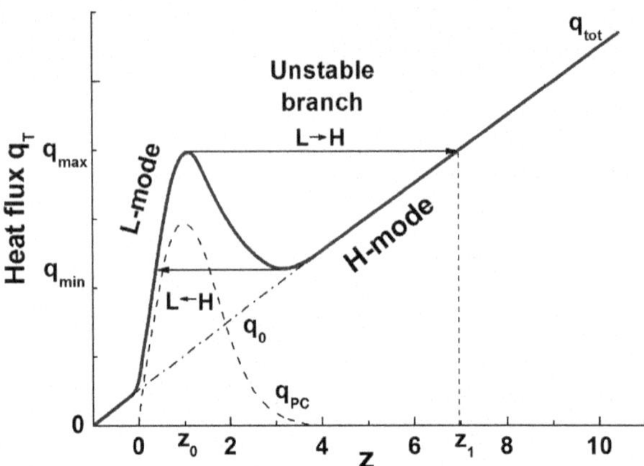

Fig. 6.1 The dependence of the heat fluxes q_{PC}, q_0 and $q_{tot}=q_{PC}+q_0$ on z, which is the deviation of the temperature profile from the critical one. When $q_{tot}>q_{max}$ the transition occurs from the L-mode branch to the H-mode branch. At $q_{tot}<q_{min}$ the inverse transition occurs from the H-mode branch to the L-mode branch

Consider the expression for the heat flux (6.4) in terms of its dependence on the temperature gradient. For simplicity we omit the indices e, i, and the convective term, and use the symbol q_{PC} and q_0 for the first and second terms in the flux (6.4)

$$q_{PC} = -\kappa T\left(\frac{T'}{T}-\frac{T_c'}{T_c}\right)H\left[-\left(\frac{T'}{T}-\frac{T_c'}{T_c}\right)\right]F, \tag{6.9}$$

$$q_0 = -\kappa^0 \frac{\partial T}{\partial \rho} \tag{6.10}$$

We also assume that the density profile coincides with the canonical density profile. Then

$$z_p = z_T = \left(\frac{\rho_{\max}}{\rho}\right)a\left(\frac{T'}{T}-\frac{T_c'}{T_c}\right). \tag{6.11}$$

For the heat flux we now have

$$q_{tot} = q_{PC} + q_0 = \kappa\frac{T}{a}\frac{\rho}{\rho_{\max}} z\exp\left(-\frac{z^2}{2z_0^2}\right)H(z)+(C_1 z + C_2), \tag{6.12}$$

where $z=-z_T$, and the coefficients $C_1=\kappa^0\rho T/(a\rho_{max})$ and $C_2=-\kappa^0 T_c{}'T/T_c$ do not depend on the temperature gradient. When $z>0$, $H(z)=1$. Figure 6.1, for example, shows the dependence of the fluxes q_{tot}, q_{PC} and q_0 on the magnitude of the deviation z at the edge of the plasma cross section $\rho=\rho_{max}$ [2]. The flux q_{tot} is a non monotonic function of z with a maximum near the point $z=z_0$ and a minimum at $z>z_0$. Between the maximum and minimum of q_{tot} the derivative dq_{tot}/dz is negative, and the corresponding quasi steady-state solution of the heat Eq. (4.2) is unstable. At low total heating power, $z<z_0$, the flux q_{tot} is less than $q(z_0)=q_{max}$ and the heat

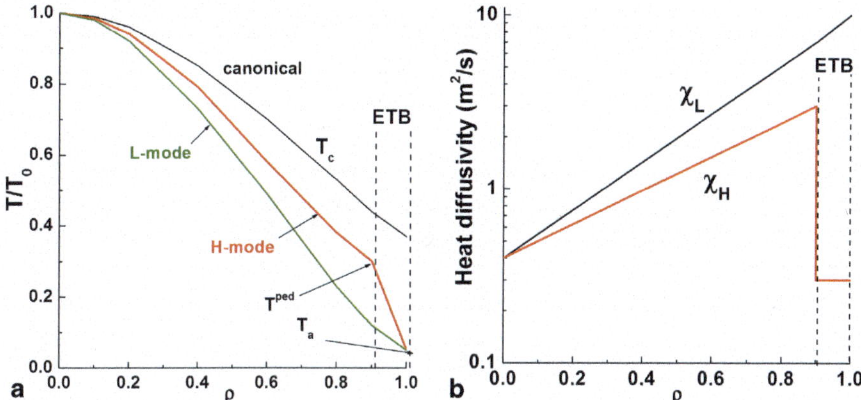

Fig. 6.2 Typical behavior of the normalized profiles of temperature (**a**) and heat diffusivities (**b**), in L- and H-modes: T_c is the canonical temperature profile, ETB is the external transport barrier

flux q_{tot} evolves along the left branch corresponding to the L-mode. If the total deposited power P_{tot} increases and becomes equal to the threshold power P_{thr}, the flux q_{tot} reaches q_{max}. At this point a bifurcation occurs and the solution jumps from one branch to another; the deviation z increases up to $z_1 >> z_0$ and the temperature gradient near the point $\rho = \rho_{max}$ increases dramatically. This bifurcation corresponds to the transition to the H-mode and the formation of the temperature pedestal at the edge of plasma.

This model includes a hysteresis with respect to the H–L and L–H transitions. When the heating power is reduced, a back transition from H-mode to L-mode occurs at a power lower than the threshold power ($P_{tot} < P_{thr}$) and the total flux jumps to the point $q_{tot} = q_{min}$. A similar process takes place also during the formation of an internal transport barrier.

6.3 Approximate Analytical Criterion for the L–H transition

We have seen above that during the L–H transition an edge transport barrier is created. The qualitative behaviour of the temperature profiles and the effective heat diffusivities (5.9) in the L- and H-modes is shown in Fig. 6.2. Here T_L and T_H are the temperature profiles in L- and H-mode, T_c is the canonical temperature profile, T^{ped} and T_a are the pedestal and boundary temperatures, χ_L and χ_H are the effective heat diffusivities in L- and H-mode. In L-mode there is a large difference between T_L and T_c in the outer part of plasma as the profile $T_L(\rho)$ is more peaked than the canonical one $T_c(\rho)$ and the boundary temperature $T(\rho_{max}) = T_a$ in the experiment is low. The high value of $|z_T|$ in the external part of plasma leads to large values of flux q_{PC} and the effective heat diffusivity χ_L. At the L–H transition a narrow transport barrier is formed in the vicinity of the edge (associated with a forgetting zone for this ETB), but in the plasma core the temperature profile $T(\rho)$ is closer to the canonical profile $T_c(\rho)$ and the deviation $|z_T|$ decreases. As a result, the effective heat diffusivity decreases both inside and outside of the barrier, in other words, over the whole plasma cross section. Inside the barrier the value of χ_H is 5–50 times smaller than the value of χ_L, and $q_{PC}=0$. The pedestal temperature T^{ped} varies widely depending on discharge conditions, from T^{ped} ~150 eV at MAST up to 7 keV at JET. In the plasma core the value for the ratio χ_L/χ_H is typically between 1.5 and 2.

Since the transport barrier is formed at the outermost part of the plasma, the condition for its formation is as follows:

$$\left|z_p(\rho_{max})\right| > z_0. \tag{6.13}$$

We put here $z_{0e}(\rho_{max}) = z_{0i}(\rho_{max}) = z_0$ and omitted the indices e, i, since at the boundary of plasma the electrons and ions are strongly linked, $T_e \approx T_i = T$, and the transport barrier is seen in all transport channels. Using the definition of z_{pk} (6.1), the condition (6.13) can be rewritten as:

$$-a\frac{T'}{T} > z_0 + a\left(\frac{n'}{n} - \frac{p_c'}{p_c}\right). \tag{6.14}$$

At the plasma edge the temperature profile in L-mode strongly differs from the canonical profile $\left|\frac{T'}{T}\right| \gg \left|\frac{T_c'}{T_c}\right|$, see Fig. 6.2a and therefore the heat flux (4.8) can be approximated by:

$$q_{tot} = -2\kappa T'(\rho = \rho_{max}) = P_{tot}/S. \tag{6.15}$$

Here q_{tot} is the heat flux per unit area, P_{tot} is the total power deposited into the plasma, S is the surface of plasma area. Solving (6.15) and substituting the result into (6.14), we obtain the condition for the transition to the H-mode

$$P_{tot} > P_{thr}, \tag{6.16}$$

where

$$P_{thr} = 2\kappa T_S\left(\frac{S}{a}\right)\left[z_0 + a\left(\frac{n'}{n} - \frac{p_c'}{p_c}\right)_S\right]. \tag{6.17}$$

The lower subscript "S" indicates that all the values in the right hand side of (6.17) must be taken in L-mode at the plasma edge $\rho=\rho_{max}$. In practical units, the formula (6.17) has the form:

$$P_{thr} = 0.0032\kappa T_S\left(\frac{S}{a}\right)\left[z_0 + a\left(\frac{n'}{n} - \frac{p_c'}{p_c}\right)_S\right], \tag{6.18}$$

where P_{thr} is in MW, κ in 10^{19} m^{-1} s^{-1}, T in keV, S in m^2 and a in m. The comparison of (6.18) with experimental data from different devices shows [2] that

$$z_0 = 8 - 10 \tag{6.19}$$

Since the value of κ is defined by (4.11) and has no radial dependence, then

$$P_{thr} \sim \bar{n} q(0.5\rho_{max}) \frac{q_{cyl}}{B} T(0.25\rho_{max})^{1/2} \left(\frac{3}{R_0}\right)^{1/4} T_S \left(\frac{S}{a}\right) \left[z_0 + a\left(\frac{n'}{n} - \frac{p_c'}{p_c}\right)_S \right]. \tag{6.20}$$

Formula (6.20) is consistent with the ITER scaling [3, 7]

$$P_{thr} = 0.7 n_{20}^{0.94} B^{0.8} R_0^{2.12}. \tag{6.21}$$

as the exponents of the density in (6.20) and (6.21) are almost identical, the factor $q(\rho_{max}/2)$ in (6.20) increases with increasing magnetic field, the multiplier q_{cyl}/B is independent of the magnetic field and the ratio S/a rapidly increases with the geometrical dimensions of the device. In (6.21) n_{20} is the average density in 10^{20} m^{-3}, and B is the toroidal magnetic field in T. However, note that formula (6.20) beside this also shows a dependence on the edge temperature and edge density gradient. With the decrease of boundary temperature T_S and/or with the increase of the density gradient the threshold power P_{thr} decreases as $n'/n < 0$.

6.4 Estimates of Transport Barrier Parameters in H-Mode

In the H-mode the plasma cross-section can be divided into two regions:

(I) core plasma, $0<\rho<\rho_{max}-\Delta$, where $F_{e,i} \approx 1$,
(II) the zone of the external transport barrier, $\rho_{max}-\Delta<\rho<\rho_{max}$, where $F_{e,i} \ll 1$

and Δ indicates the width of the transport barrier. The ion-electron exchange is large at the plasma edge, so here $T_e \approx T_i = T$. To simplify our estimation we also assume that $\kappa_e^{PC} = \kappa_i^{PC} = \kappa$, and that the electron and ion heat fluxes are equal. In this case, the heat flux q_T (6.4) in the electron or ion channel in the edge part of region (I), i.e near the transport barrier, can be written as follows:

$$q_T = -\kappa T F \frac{d}{d\rho} \ln\left(\frac{T}{T_c}\right) = \frac{1}{2}\frac{P_{tot}}{S}, \quad T = T^{ped}. \tag{6.22}$$

Here P_{tot} is the total power deposited into the plasma, S is plasma surface and the factor 1/2 takes into account that only half of the total power is deposited into electrons or ions. Inside the transport barrier we have

$$q_T^+ = -\kappa^0 \frac{dT}{d\rho} = \frac{1}{2}\frac{P_{tot}}{S}, \quad q_T^+ = q_T^- = q_T. \tag{6.23}$$

At the point near the boundary of areas (I) and (II), where the absolute values of the first and of the second terms in (6.4) are comparable, we have

$$\kappa \exp\left(-\frac{z_p^2}{2z_0^2}\right) = \kappa^0. \tag{6.24}$$

Hence, we determine the value of z_p:

$$z_p = z_T + z_n = -z_0\left[2\ln\left(\frac{\kappa}{\kappa^0}\right)\right]^{1/2}. \tag{6.25}$$

From this formula, we find

$$z_T = -\zeta, \quad \zeta = z_0\left[2\ln\left(\frac{\kappa}{\kappa^0}\right)\right]^{1/2} + z_n, \quad z_n = a\left(\frac{n'}{n} - \frac{n_c'}{n_c}\right) \tag{6.26}$$

or

$$a\frac{d}{d\rho}\ln\left(\frac{T}{T_c}\right) = -\zeta. \tag{6.27}$$

Now we substitute (6.27) in (6.22). As a result, we obtain

$$T^{ped} = \frac{1}{2}\frac{P_{tot}a}{S\kappa^0\zeta}. \tag{6.28}$$

Here we have used (6.23). In practical units, this formula has the form

$$T^{ped} = 625\frac{P_{tot}a}{2S\kappa^0}\left[z_0\left(2\ln\left(\frac{\kappa}{\kappa^0}\right)\right)^{1/2} + z_n\right]^{-1} \tag{6.29}$$

Here, κ and κ^0 are in 10^{19} m^{-1} s^{-1}, a in m, S in m^2, P_{tot} in MW, T^{ped} in keV, z_0 and z_n are dimensionless. To make an estimate in what follows, we neglect the influence of the density profile on the value of the pedestal temperature and use the simpler formula

$$T^{ped} = 625\frac{P_{tot}a}{4S\kappa^0 z_0}\left[\ln\left(\frac{\kappa}{\kappa^0}\right)\right]^{-1/2} \tag{6.30}$$

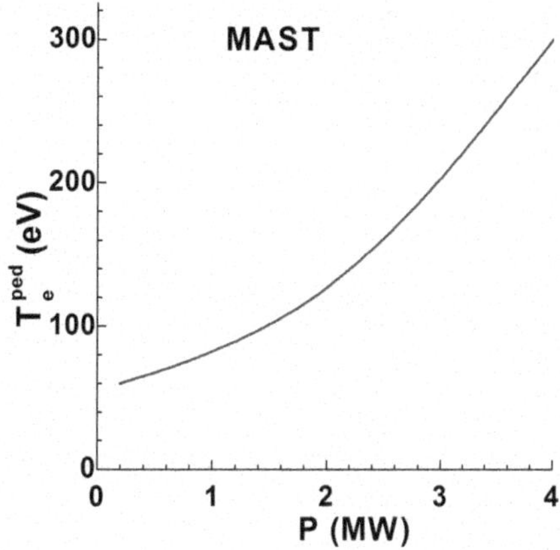

Fig. 6.3 The diagram of the experimental dependence of the electron temperature pedestal T_e^{ped} on the deposited power in MAST

An increase in T^{ped} with increasing heating power was observed in a number of experiments. Figure 6.3 shows the dependence of T^{ped} on $P_{SOL}=P_{tot}-P_{rad}$, found in MAST experiments [8]. Here P_{rad} is the radiated power. It can be seen that T^{ped} increases almost linearly with increasing heating power. Integrating (6.23) over the transport barrier, we obtain

$$\kappa^0 T^{ped} \approx \frac{1}{2}\Delta\frac{P_{tot}}{S}. \tag{6.31}$$

Here we have taken into account that the $T_S/T^{ped} \ll 1$. Using (6.28), we find

$$\frac{\Delta}{a} = \frac{1}{\zeta}. \tag{6.32}$$

Values for the heat conductivity coefficient κ^0 for different plasma regimes in JET will be discussed below.

Experimentally, the pressure or temperature gradient inside the barrier is sometimes limited to a certain value. This can be caused by e.g. ballooning instabilities. If the absolute value of the temperature gradient is limited by the value $T_b' > 0$, then two cases are possible:

(a) The absolute value of the temperature gradient within the barrier is less than T_b', i.e.

$$\frac{q_T^+}{\kappa^0} < T_b'. \tag{6.33}$$

In this case formulas (6.28, 6.32) are unchanged.

(b) In the other case, i.e.

$$\frac{q_T^+}{\kappa^0} > T_b'. \tag{6.34}$$

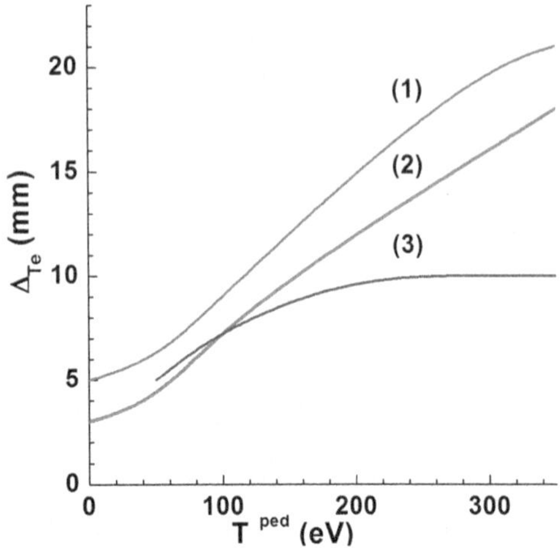

Fig. 6.4 The diagram of the dependences of the width of the electron temperature ETB Δ_{Te} on the temperature pedestal value for three types of MAST discharges

(6.32) does not hold, but the formula (6.28) for T^{ped} is unchanged. Therefore, for the width of the barrier one should use the obvious expression

$$\frac{\Delta}{a} = \frac{T^{ped}}{aT_b'} = \frac{P_{tot}}{2S\kappa^0\zeta T_b'}. \tag{6.35}$$

This shows that the width of the transport barrier is proportional to the heating power. This was indeed seen on MAST [8] and Alcator C-Mod [9]. Figure 6.4 shows the experimentally found dependence of the barrier width in the electron temperature Δ_{Te} versus pedestal temperature T_e^{ped} at different plasma cross sections and for different discharge regimes in MAST. It is seen that in discharges (1) and (2) the value of Δ_{Te} rises with the increasing pedestal temperature. Since T_e^{ped} increases with increasing heating power (Fig. 6.3), Δ_{Te} also increases with increasing heating power. This is in agreement with formula (6.35). However, it is also clear in Fig. 6.4 that in discharge (3) the value of Δ_{Te} saturates with increasing T_e^{ped}. This can be interpreted as a transition to a regime described by Eq. (6.33). The width of the barrier is then described by (6.32), which contains no dependence on the heating power.

6.5 Transport Coefficients in the Nonlinear Model

Discharges in the H-mode are typically not in steady state. At the plasma edge where the pressure gradient is large, periodic oscillations occur, which can be described as follows. First, the energy stored in the plasma edge is ejected during a time interval of the order of tens or hundreds microseconds. Subsequently during a time interval of the order of tens milliseconds, the temperature and density profiles are reestablished, and the cycle can start again. Typically, the amount of energy that is ejected from the plasma edge is a few percent of the total stored plasma energy.

In what follows, we will call such relaxation oscillations, ELMs (Edge Localized Modes).

The frequency depends on the size of the device and the plasma scenario. In plasmas with so-called Type III ELMs the oscillation period is short, and the amount of energy ejected is small. This is the type of ELMs that is foreseen for future fusion reactors. In regimes with Type I ELMs the oscillation period is much longer and the amount of energy ejected can reach up to 3–4 % of total stored plasma energy. This type of ELMs is should be avoided for a future reactor, because the amount of energy ejected can damage the first wall and/or the divertor. To control the type of ELMs experimentally is very much like an art. Most transitions between different types of ELMs occur spontaneously and their causes remain unclear. Sometimes ELMs suddenly dissappear, and then a so-called ELM-free H-mode is established. In many devices special discharge scenarios have been established with ELM-free periods during a short time.

The current theoretical understanding of the formation of ELMs can be summarized as follows [10]. The pressure gradient in the ETB is limited by MHD ballooning modes. When the pressure gradient exceeds a certain threshold value, such a MHD ballooning mode appears, thereby stopping further growth of the pressure gradient. However, the existence of ELMs is determined by another instability: the so-called peeling instability. This type of instability becomes unstable when the current density at the plasma edge exceeds a certain critical (fairly large) current density. Thus, the description of ELM oscillations is as follows. During the ejection of energy and particles to the wall, the edge current is also ejected, therefore, the current density at the plasma edge decreases. Pedestal density and temperature then recover to some intermediate level very quickly, but not the edge current density. The ELM frequency is thus determined by the current diffusion at the outer plasma edge and a next ELM will only occur when the current density at the edge again exceeds the critical current density.

Experimentally, the L–H transition is characterized in hydrogen (or deuterium) plasmas by the intensity at the plasma edge of the Balmer-α and/or Balmer-β lines H_α and/or H_β of hydrogen (and similarly, D_α and/or D_β lines in the case of deuterium plasma). The intensity of these lines is proportional to the product of the density of the cold hydrogen (or deuterium) neutrals and the electron density, but this product is highly peaked at the plasma edge. In the L-mode the lines are quite intense and somewhat spread out in radial direction. At the transition to H-mode the intensity decreases, and is localized in a narrower radial zone, since most of the neutrals are absorbed inside the transport barrier. The decrease in the intensity of the radiation is determined by the decrease in the density of the cold neutrals due to the improved particle confinement in the plasma, and, as a result, in the decreased flux of neutrals reflected from the wall. In presence of an ELM, a sharp intensity peak is observed with a duration of the order of tens or hundreds microseconds. This peak marks the first short phase of the ELM oscillation. The distance between the peaks determines the period of these oscillations. Figure 6.5 shows, as an example, the time evolution of the intensity of the D_α radiation for JET discharge #79698. All the features, which we have just discussed above, are clearly visible in this diagram.

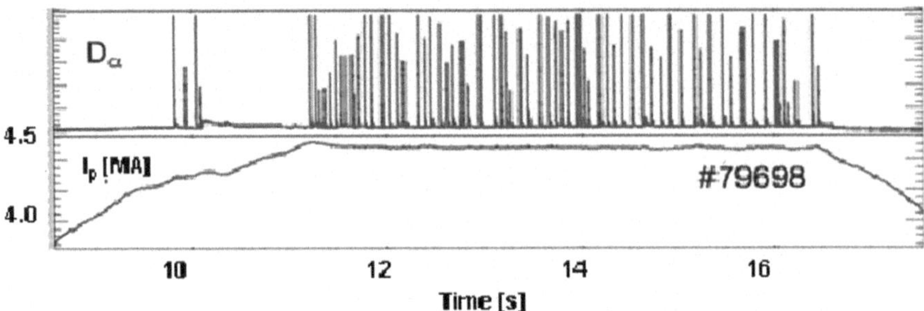

Fig. 6.5 The dependences of the plasma current and intensity of line radiation D_α on time for the JET discharge #79698. The moment of the L–H transition, a period of the ELM-free discharge and further highlights of the ELMs appearance are visible

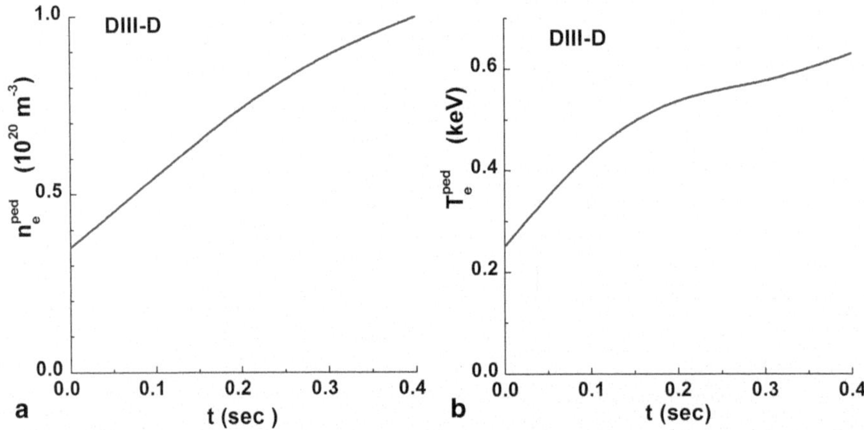

Fig. 6.6 The scheme of the growth of the pedestal density (**a**) and of the pedestal temperature (**b**) after the transition from the ELMs regime to the ELM-free regime in DIII-D. Here *t* is time after the last ELM. The ELMs period was of the order of a few tens milliseconds

How can we formulate a transport model for such a non steady state plasma behavior without a detailed description of each ELM? Two important observations come to rescue. First, the ELM period, even for Type I ELMs, is much smaller than the energy confinement time. Therefore, the impact of the ELMs on the transport model can be averaged over time. Second, the recovery of the pedestals as a function of time has a specific character that helps to build a diffusive transport model as explained below.

Figure 6.6 shows the time evolution of the density and temperature pedestals during the transition from an ELMy regime to an ELM-free regime in DIII-D [11, 12]. It is clear that immediately after the ELM crash, during a short time interval of the order of 10 ms or less, a rapid partial recovery of the pedestals to an intermediate value took place. Then the growth rate of the pedestals is strongly reduced and the subsequent pedestal recovery takes place during a time interval of the order of one to a few hundreds of milliseconds. Since the ELM period does not exceed a few tens of milliseconds, the intermediate values of the pedestals indicated above, are a good approximation of the average values over the pedestal period. Thus, the concept of the "pedestal during the ELMs" is well defined, and can be used in the construction of the transport model. Pedestal values for Type I or Type III ELMs are not very different, as during the second phase of the ELM the recovery rate is rather slow.

The time-averaged transport caused by the ELMs is described by the transport coefficients $\chi^0_{e,i}$ and D^0 in the region where the ETB is localized, and characterize the loss of energy and particles at ELM crashes, and also the incomplete recovery of the pedestals between ELMs. In ELM-free regimes the plasma reaches steady state, so the coefficients $\chi^0_{e,i}$ and D^0 describe the real diffusion transport processes within the ETB. Calculations have shown that in ELM-free plasma phases these coefficients should be 8–10 times smaller than for ELMy plasmas.

In the non-linear model for the H-mode the transport coefficients for the temperature, plasma density and toroidal velocity, κ^{PC}_k, D^{PC} and χ^{PC}_ω are considered the same as in the linear model (5.12, 5.16, 5.17, 5.18, 5.19), but the coefficients κ^0_k (or χ^0_k), D^0 and χ^0_ω, determining the transport inside the ETB, have to be found. To this end, we need to solve the inverse problem of finding the coefficients in the set of Eqs. (5.1, 5.2, 5.4) using a sufficiently representative experimental database. In the non-linear model the boundary conditions for the electron and ion temperatures and plasma density are taken at the separatrix, so the gradient region and zone where the ETB is localized are included in the statement of the problem. The presence of a pedestal is not clearly seen in the toroidal rotation profile, so the value of χ^0_ω is uncertain so far.

The following example illustrates how the coefficients χ^0_k and D^0 are determined using experimental data. ELMy discharges from JET were used, as described in Table 6.1 [13]. The problem was solved by the selection method. Simulating the set of discharges from Table 6.1, we find the parameters χ^0_e, χ^0_i and D^0, which minimize the deviation between the calculated and experimental temperature and density profiles. For the RMS deviations of the temperature profile, we use the following expression:

$$d_2 T = \left\{ \frac{1}{N} \sum_k^N \left[\frac{T_k^{calc} - T_k^{exp}}{T_k^{exp}} \right]^2 \right\}^{1/2}. \tag{6.36}$$

Rounding the values found for the coefficients χ^0_k and D^0 we obtain for ELMy discharges

$$\chi^0_i = 0.46 \frac{P_{tot}}{\langle n \rangle I_p}, \quad \chi^0_e = 1.5 \chi^0_i \sqrt{T_e}. \tag{6.37}$$

For the particle diffusion coefficient we found the much simpler expression:

$$D^0 = 0.05. \tag{6.38}$$

Table 6.1 The parameters of the chosen JET shots

No	Dicharge	t	I	$\bar{n}$	P_{NB},	q_a	$T_n(\rho_{max})$
		s	MA	10^{19}m^{-3}	MW		
1	61103	15	2.75	6.4	15	3.13	0.64
2	61138	20	2.5	10.4	14	3.66	0.67
3	61174	22	2.35	5.6	12	2.91	0.58
4	61132	25	2.35	2.3	2.5	3.16	0.35
5	61097	23	2	4.8	8	2.96	0.6
6	61366	26	1.5	2.9	14	4.27	0.72
7	61543	23	1.5	4.14	14	4.52	0.64
8	61520		1.4	3.1	14	6.3	–
9	61526	22	1	2.4	7.9	6.3	0.75
10	61236		1	2.3	12.5	11	–
11	62094	16	2.75	3.1	13.4	3.04	0.4
12	62093	16	2.75	4.1	16.2	3.06	0.5
13	69463	22.8	1.7	5.3	8.2	5.2	0.65
14	69463	26	1.7	5	14.5	5.3	0.55
15	78032	13	2.5	5.74	15.2	4.5	0.58

In (6.37 − 6.38) n, I_p, P_{tot}, R_0, $T_{e,i}$, χ and D are expressed in MA, MW, m, keV, m²/s and the symbol <…> indicates volume averaging.

However, the inverse problem for the determination of D^0 is severely ill-posed. When the value of D^0 changes by a factor of 5, the value of the pedestal density changes by less than 10 %.

The typical behavior of the ion heat diffusivity in JET ELMy H-mode plasmas is shown in Fig. 6.7, where the profiles of the effective ion heat diffusivity $\chi_i^{eff} = -q_i/(n\partial T_i/\partial\rho)$ and the profiles of χ_i^0 are plotted.

To describe JET ELM-free H-mode discharges the values of the coefficients χ_i^0 and χ_e^0 should be 8–10 times smaller than those given by (6.37). Only then we find pedestal values close to the experimental ones, which are 3–4 times greater than in ELMy discharges. We will show below two examples for JET ELM-free H-mode discharges. Formulas (6.37–6.38) are not yet universal because experiments in a particular device cannot reveal the dependence on the geometry of the plasma. E.g. the analysis of experiments in MAST has shown that for ELMy discharges the coefficients χ_i^0 and χ_e^0 have to be 2 times larger than those determined in formulas (6.37).

6.6 General Remarks on the Second Critical Gradient

The first critical gradient for the temperature $-RT_c'/T_c$ is determined in our model with the help of the canonical pressure profile $RT_c'/T_c = \frac{2}{3}Rp_c'/p_c$ (1.77). The differential Eq. (1.98) is valid for the canonical profiles together with the boundary conditions (1.99)–(1.101). Therefore the first critical gradient is defined by a closed theory. For the second critical gradient we have yet no theory. But we have seen that a relatively simple phenomenological model with a forgetting factor allows the

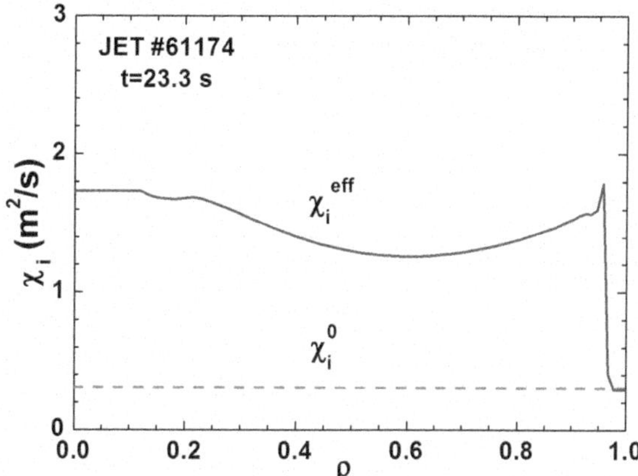

Fig. 6.7 Profiles of the effective ion heat diffusivity χ_i^{eff} and background heat diffusivity χ_i^0 which determines the heat transport inside the ETB

description of the complex physics of the external transport barrier. Therefore, let us try now a similar approach for the internal barrier.

The discussion that now follows is divided in two parts. First, we consider the problem of determining two functions $z_{0e}(\rho)$ and $z_{0i}(\rho)$ from the comparison of the calculations with experiment. To do this, we represent these functions in parametric form, and then define the parameters, by the modelling the discharges with improved confinement and minimizing the RMS deviation. Then, knowing the spatial behavior of $z_{0e}(\rho)$ and $z_{0i}(\rho)$, we compare them with the spatial dependencies of safety factor $q(\rho)$ and shear of the poloidal magnetic field $s=\frac{\rho}{q}\frac{\partial q}{\partial \rho}$.

Recall the definition of the functions $z_{0e}(\rho)$ and $z_{0i}(\rho)$ at the outer plasma edge. Since the transport barrier is narrow and is located at the edge, it can be described in the model by the values $z_{0e}(\rho_{max})$ and $z_{0i}(\rho_{max})$ only. Since the electrons and ions are strongly coupled at the plasma edge it is naturally to assume that $z_{0e}(\rho_{max})=z_{0i}(\rho_{max})$. For the sake of brevity we denoted this value by z_0. Thus, the L–H transition in the model is determined by only one number that we need to find, comparing the calculations with experiment. This comparison was carried out in [2], where it was shown that for DIII-D $z_0=8-9$ and for JET $z_0=6-7$. Our calculations for MAST, a low aspect ratio device with $A\sim 1.5$, have shown that in this case $z_0=5-6$. Thus, the value of z_0 is apparently in the range $6<z_0<9$ and slightly increasing with increasing aspect ratio.

The behaviour of $z_{0e}(\rho)$ and $z_{0i}(\rho)$ in the core and gradient zone of the plasma can be determined by the analysis of discharges with an ITB. This was done in our papers [4, 14], where we show that the value of these functions inside the plasma are much smaller than the value of z_0, especially for the ions. The results of the analysis can be conveniently represented in the form of a piecewise linear function (with $k=i, e$):

$$z_{0k}(\rho)=\begin{cases}\alpha_{k1} \text{ at } 0<\rho<\rho_1\\ \alpha_{k1}+\dfrac{(\alpha_{k2}-\alpha_{k1})(\rho-\rho_1)}{(\rho_2-\rho_1)} \text{ at } \rho_1<\rho<\rho_2\\ \alpha_{k2} \text{ at } \rho_2<\rho<\rho_{\max},\end{cases} \tag{6.39}$$

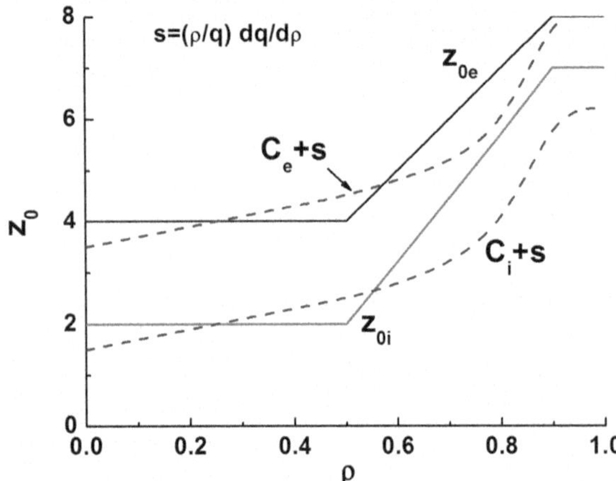

Fig. 6.8 The threshold of formation of internal and external transport barriers for electrons and ions, z_{0e} and z_{0i}, (piecewise-linear *solid lines*) and the approximation of thresholds with curves of the form $C_{ei}+s$, where s is a shear of the poloidal magnetic field. The threshold of the ETB formation (H-mode) at $\rho\sim 1$ is almost identical at both channels

where

$$z_{0e}(0)=\alpha_{e1}=4-5,\ z_{0i}(0)=\alpha_{i1}=2-3,\ \alpha_{e2}=\alpha_{i2}=z_0=6-8,$$

$$\frac{\rho_1}{\rho_{max}}=0.5-0.6,\ \frac{\rho_2}{\rho_{max}}=0.8-0.9. \tag{6.40}$$

The profiles of the functions $z_{0e}(\rho)$ and $z_{0i}(\rho)$ for JET are shown in Fig. 6.8.

Let us now compare the conditions for the formation of internal and external transport barriers. The general conditions for the formation of the barrier (6.2) can be rewritten as

$$a\frac{\rho_{max}}{\rho}\left|\frac{p_k'}{p_k}-\frac{p_c'}{p_c}\right|>z_{0k}(\rho) \tag{6.41}$$

It is evident that for the left side of (6.41) to be sufficiently large to satisfy the inequality, the following conditions must be satisfied:

1. The power deposition profile should be strongly peaked in order to obtain a sufficiently large pressure gradient.
2. The power deposited inside the ITB in the plasma core, must be larger than some threshold power.
3. Increasing the density gradient in the ITB region reduces the power threshold.

For the L–H transition and the appearance of the ETB the first condition is not necessary. Fulfilling condition (2) for the L–H transition is easier because the pressure p_k at the plasma edge is low and fixed and therefore the pressure gradient in the L-mode could be sufficiently large. However, the threshold value for the pressure gradient at the plasma edge, z_0, is 2–3 times larger than the values z_{0k} in the gradient zone (6.39). Therefore, the presence of an ITB does not imply the simultaneous presence of an ETB and vice versa. There exist, of course, discharge types in which both an ITB and ETB appear.

We now consider the physical mechanisms that determine the behavior of the functions $z_{0k}(\rho)$ ($k=e, i$) at the right-hand side of (6.41). It is generally accepted that the following three effects influence the formation of an ITB:

1. Toroidal and poloidal plasma rotation shear $s_\upsilon = \frac{\rho}{\upsilon}\frac{\partial \upsilon}{\partial \rho}$ in a narrow zone within the barrier.
2. Negative or low positive magnetic shear $s = \frac{\rho}{q}\frac{\partial q}{\partial \rho}$ in the area of formation of the barrier.
3. The existence of principal resonant surfaces $q=m/n$ with a sufficiently low m and n in the area of formation of the barrier.

To determine the link between rotation and transport we consider the force balance equation for the ions, omitting both the viscosity and quadratic velocity terms

$$\frac{dp_i}{d\rho} = enE_\rho + \frac{en}{c}(\upsilon_{i\theta}B_\varphi - \upsilon_{i\varphi}B_\theta) \tag{6.42}$$

where $p_i=nT_i$ is the ion pressure, E_ρ is the radial electric field, B_θ and B_φ are the poloidal and toroidal components of the magnetic field and $\upsilon_{i\theta}$ and $\upsilon_{i\varphi}$ are the poloidal and toroidal ion velocities. The Eq. (6.42) links the four unknown functions (p_i, E_ρ, $\upsilon_{i\theta}$ and $\upsilon_{i\varphi}$), and therefore, can only be used for a qualitative assessment. This equation shows that a large pressure gradient inside the barrier can lead to a large value for the radial electric field. A large pressure gradient can also create high poloidal and toroidal velocities $\upsilon_{i\theta}$ and $\upsilon_{i\varphi}$ for the plasma ions. In turn, the high radial electric field may cause a drift of the plasma with velocity $\mathbf{\upsilon}_\mathrm{d} = c[\mathbf{E}\times\mathbf{B}]/B^2$, resulting in a toroidal and/or poloidal rotation with high shear. This sheared rotation of the plasma is usually taken as the main reason for the suppression of turbulence and reducing anomalous transport.

In this context, a lot of experimental work is devoted to clarify the sequence of events during the L–H transition. The main question, which remains still unanswered, is what happens first: the L–H transition or the increase in plasma rotation? The strong link between the L–H transition and the radial electric field is confirmed by experiments where a charged electrode is inserted into the plasma, up to 1–2 cm inside the last closed magnetic surface, in so-called biasing experiments. Such experiments reproduce effects which are very similar to those observed during the L–H transition. If the plasma conditions are close to those for an L–H transition, one can also try other methods to trigger the L–H transition; e.g. a small puff of hydrogen gas (TUMAN-3) [15], or a small shift of the plasma (T-10) [16].

Concering the modeling of the L−H transition with the canonical profiles transport model, the following observation is important. Experimentally no precursor has been observed prior to the transition. All changes in the plasma parameters during the transition seem to take place almost simultaneously. The experimental criterion for the transition $P_{tot}>P_{thr}$ contains only one parameter (P_{tot}) and is similar in nature as criterion (6.41). Thus, the canonical profiles transport model may describe the L–H transition consistently, although it does not contain a detailed description of the physical processes taking place during an L–H transition.

Some remarks concerning the formation of ITBs. Already in the early years of ITB research (1993–1995) it was shown that the criterion for the formation of an ITB depends on the profile of $q(\rho)$ and the poloidal magnetic field shear s. At low heating powers, an ITB is formed if $s<0$ in the core plasma. Under these conditions, generally a «strong» ITB of the «box» type is formed, with a small width and very low transport coefficients inside the barrier. Usually the "strength" of the barrier is characterized by the value of $R_0/L_T=-R_0T'/T$ inside the barrier. For a strong barrier $R_0/L_T=30-40$.

For positive but low shear ($0<s<0.2$) an ITB can also be formed in the core plasma, but this then requires more heating power than in the previous case. Under these conditions, generally a "weak" ITB is formed with a large width but lower characteristic value of $R_0/L_T=20-25$. After its formation, the transport barrier can move to the plasma edge, but generally, it stops in the vicinity of the point $\rho/\rho_{max}\sim 0.65-0.7$. Measurements of the plasma rotation velocities near the ITB have shown that there are zonal plasma flows in this region.

We come to the following conclusions to summarize it all:

1. At the plasma edge the magnetic shear is great. For plasmas with a high aspect ratio $A=5$ and circular cross-section (T-10) the value of s at the edge is about $s\approx 2$. At $A\approx 1.5$, and elongation $k\sim 2$ (MAST) the value of s at the plasma edge is $s=5$–6. With such a high shear the details of the presence of resonant surfaces seem to play a secondary role. The radial electric field and zonal near-edge plasma flow play a major role in the formation of an ETB. In this case, the CPTM, developed above, can reasonably describe the L–H transition, and with the help of (6.42), knowing $\frac{\partial p}{\partial \rho}$, to estimate the radial electric field. To close the model we need only to specify one parameter z_0 by comparing the calculations with experiment. This parameter can depend on the geometry of the plasma.
2. In the plasma core the formation and position of an ITB depends on the magnitude and sign of the shear s and on the positions of resonant surfaces r_{s1}, $r_{s3/2}$, r_{s2}, …, as well as on the value and profile of the heating power. In our model the criterion for the formation of an ITB has the form (6.41). The magnitude and the profile of the heating power are reflected in the left-hand side of (6.41). Other dependencies are reflected in the right-hand side, e.g. via the choice of the functions $z_{0k}(\rho)$ ($k=e, i$) over the entire plasma cross section. It is useful to compare the behaviour of the functions $z_{0k}(\rho)$ with the behaviour of shear of the poloidal magnetic field s. In Fig. 6.8 we also show the typical behaviour of the functions $s+$ const at low positive shear in the plasma core (dotted lines). One sees clearly that the regions coincide where both functions $z_{0k}(\rho)$ and $s(\rho)$ show a steep rise. Thus the function $z_{0k}(\rho)$ can be approximately written in the form

$$z_{0k}(\rho)\approx C_k+C_{1k}s(\rho),\quad (k=e,i). \tag{6.43}$$

Table 6.2 The optimal values of the parameters of the function (6.43) for different devices

Device	Dicharge No	C_i	C_{1i}	C_e	C_{1e}	χ_i^{ITB} m²/s	Δ^{ITB}/a Model
JET	40847	1	2	5	2	0.24	0.2
TFTR	94607	1.5	2	4	–	0.2	0.2
DIII-D	89943	1	2	4	–	0.22	0.45
MAST	8575	1	2	>3	2	0.6	0.34

The expression (6.43) contains 4 parameters (C_k and C_{1k}, two for electrons and two for ions), which have to be determined from experiment.

Calculations [14], carried out for several devices, led to optimal values of the parameters of the function (6.43), summarized in Table 6.2, and a good approximation for the various constants is given by:

$$C_i = 1 - 1.5, \quad C_e = 4 - 5, \quad C_{1i} = C_{1e} = 2. \tag{6.44}$$

Expression (6.43), in particular, describes reasonably well the possible movement of an ITB after its formation, as the s profile depends on time. Figure 6.8 also shows that in the neighbourhood of the point $\rho/\rho_{max} \sim 0.7$ the magnitude of the shear begins to increase rapidly in radial direction, causing a stop in the movement of the ITB.

The structure of the function (6.43) does not take into account the effect of resonant surfaces on the formation of an ITB. To include this effect in the model, it is necessary to know the behaviour $q(\rho)$ over the plasma cross section. If it is known, it is possible to find the position of the principal resonances ρ_{s}. Let us introduce the conception of "gap" as an interval around the point ρ_{s}, where there are no other resonant surfaces with low m and n. The width of gap is detemined (mainly intuitive) by maximal values of m and n for MHD modes which can still play an important role in the transport. These estimates may be different for different resonant surfaces. For examples, for the surface $q_s = 1$, $m \sim n < 7 - 10$, for surface $q_s = 2$, $m < 20$, $n < 10$. It is clear that the widths of the gaps increase when the function $q(\rho)$ becomes more flat.

If the positions ρ_s and the widths of the gaps δ_s for the principal resonances are identified, then their impact on the formation of an ITB can be taken into account through the function $z_{0k}(\rho)$ as follows. Let's introduce for each gap the notion of "depth" h_s (the value has to be determined by comparing calculation and experiment) and define the function $\Pi_s = \Pi(\rho_s, \delta_s, \rho)$ as a rectangular function

$$\begin{aligned} \Pi_s &= 0 \text{ at } 0 < \rho < \rho_s \text{ and } \rho_s + \delta_s < \rho < \rho_{\max} \\ \Pi_s &= h_s \text{ at } \rho_s < \rho < \rho_s + \delta_s. \end{aligned} \tag{6.45}$$

Here s is the number of the resonant surface. Next we define the function

$$z_{0k}(\rho) \approx C_k + D_k s(\rho) - \sum_s \Pi_s. \tag{6.46}$$

The function (6.46) includes the effect of the position of the resonant surfaces, and of the magnetic shear value on the formation of the internal transport barrier. In many examples only one term is considered in the sum of Π_s as only the one corresponding to $q_s=2$ or 3/2 plays a significant role.

The inequality (6.41) can be used to estimate the distance δ_{12p} between the first and second critical pressure gradient

$$\delta_{12p} = R_0\left|\frac{p_2'}{p_2} - \frac{p_c'}{p_c}\right| > z_{0k}(\rho)A\frac{\rho}{\rho_{max}}. \tag{6.47}$$

Assuming that the plasma density profile is close to the canonical profile, we obtain an estimate for the distance between the first and second critical temperature gradient δ_{12T}

$$\delta_{12T} = R_0\left|\frac{T_2'}{T_2} - \frac{T_c'}{T_c}\right| > z_{0k}(\rho)A\frac{\rho}{\rho_{\max}}. \tag{6.48}$$

In particular, for JET ($A=3$) at the formation of the *i*-ITB at half the plasma radius ($z_{0k}=3$) we have $R_0|T_2'/T_2 - T_c'/T_c|=4.5$. According to Fig. 2.9, $-R_0 p_c'/p_c \sim 7.5$, and the first critical gradient is $-R_0 T_c'/T_c = \frac{2}{3}R_0 p_c'/p_c \sim 5$. Therefore, the second critical gradient is equal to $-R_0 T_2'/T_2 \approx 9.5$. In this case, the second critical gradient is almost two times larger than the first critical gradient.

It is useful to compare these estimates with the experimental criterion for ITB formation, obtained in JET [17]

$$\frac{\rho_S}{L_T} \geq 0.014, \quad \left[L_T = -\frac{T}{T'}, \quad \rho_S = \frac{\upsilon_S}{\omega_B}, \quad \upsilon_S = \left(\frac{2T_e}{m_i}\right)^{1/2}\right] \tag{6.49}$$

Here υ_S is the speed of sound. In practical units (ρ_S in m, B_0 in T, T_e in keV)

$$\rho_S = 0.0046\frac{\sqrt{T_e}}{B_0}. \tag{6.50}$$

Inequality (6.49) can be rewritten as follows:

$$\frac{R_0}{L_T} > \frac{3R_0 B_0}{\sqrt{T_e}}. \tag{6.51}$$

Substituting in (6.51) typical values for JET parameters ($T_e=9$ keV, $B_0=3$ T), we obtain the experimental value of the second critical gradient $-R_0 T_2'/T_2=9$. Our model gives for the second critical gradient the value of 9.5, so the difference with experiment is less than 10 %. Until now, we have assumed that the value of z_{0k} does not depend on the plasma parameters. The JET experiments indicate that z_{0k} may be proportional to $B_0/\sqrt{T_e}$.

To conclude this section we discuss the natural question why a plasma with a non-monotonic current profile (i.e. far from the canonical current profile) in general can be described reasonably well by the proposed system of Eqs. (5.1–5.4) with fluxes (5.8, 5.10, 5.11)?

However, as we have already pointed out above, *only the canonical profiles of current and pressure* (the relaxation goals) *should match.* Real profiles of these parameters can be rather different. In Eq. (5.3), which describes the evolution of the current profile, the characteristics of canonical profiles are not included. On the other hand, the actual current profile appears in the equilibrium Grad–Shafranov Eq. (2.2) and is thus taken into account in the geometry of the plasma and in the determination of the metric coefficients. Canonical profiles are introduced only into the expressions for the fluxes of heat, particles and momentum through the critical gradients. Regimes with a non-monotonic current are non steady state, and the relaxation goes in the direction of the canonical profiles. But the current profile relaxes very slowly.

6.7 Examples

Here we present the results of calculations of the transport system of Eqs. (5.1–5.4) with fluxes (6.4–6.6) for the H-mode, and for discharges with internal transport barriers. For comparison with the experiment the ITER database was used [18]. In all calculations we assumed that $z_0 = 8$.

Let's start with the modeling result for JET discharge #61174 (see Table 6.1) as shown in Fig. 6.9. The experimental and calculated time evolution of the central electron and ion temperatures is shown in Fig. 6.9a. It can be seen that the calculations reasonably reflect the time evolution of the experimental temperature behavior. However, the experimental curves are very noisy and this causes the mean square (RMS) deviation to be large, on the order of 8–10 %. The following Fig. 6.9b and c show the experimental and calculated profiles of the electron temperature and plasma density at $t = 21$ s. For the boundary conditions at the separatrix the following values were taken: $T_e(a) = 50$ eV and $n(a) = 0.5 \times 10^{19}$ m^{-3}. The electron temperature pedestal reaches 0.8 keV and the plasma density pedestal 3.5×10^{19} m^{-3} with an RMS deviation between the experimental and calculated curve less than 5–8 %. Figure 6.9d shows the calculated curves for the effective heat diffusivity of electrons and ions at time $t = 21$ s. In this case, the heat diffusivity inside the ETB is by one order of magnitude smaller than in the plasma core.

Figure 6.10a and b show the experimental and calculated profiles of electron and ion temperature at the time $t = 21$ s for JET discharge #26987 which had a low chord averaged density $\bar{n} = 3.2 \times 10^{19}\,\mathrm{m}^{-3}$ and heating power $P_{NB} = 12.8$ MW. Unfortunately, there are no experimental data for the temperature inside the ETB and for the width of the barrier. It can be seen from Fig. 6.10a and b for the another JET discharge #26087, that the calculated and experimental values for the temperature pedestals became comparable and equal to about 2 keV. The central temperatures in the calculation and experiment are also rather close.

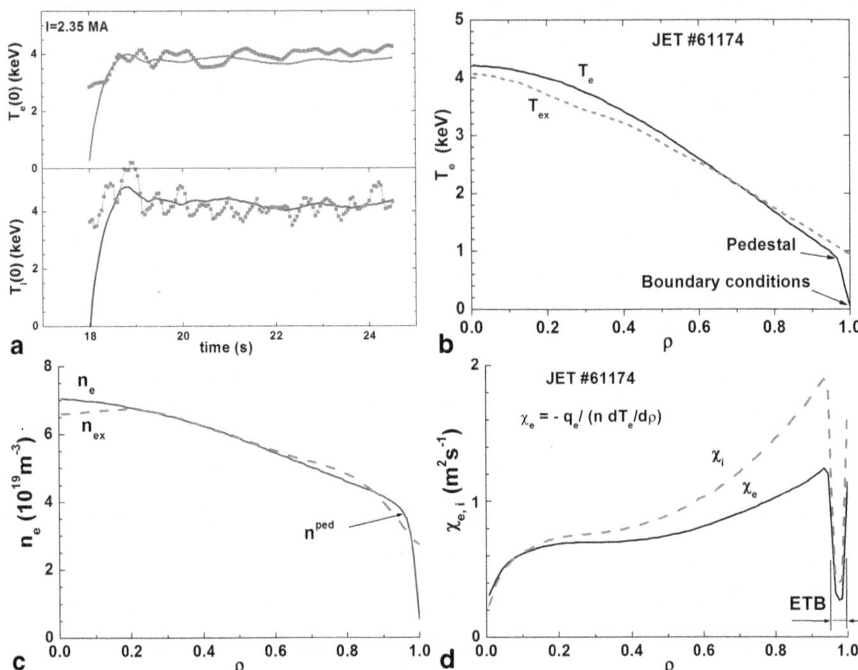

Fig. 6.9 **a** Experimental (*red squares*) and calculated (*solid curves*) behavior of the central electron (*top*) and ion (*bottom*) temperatures over time. **b** Experimental (*dotted line*) and calculated (*solid line*) profiles of the electron temperature. **c** Experimental (*dotted line*) and calculated (*solid line*) profiles of the plasma density. **d** The effective heat diffusivities of electrons χ_e and ions χ_i. The curves in (**b**), (**c**), (**d**) for the JET discharge #61174 at the time t=21 s

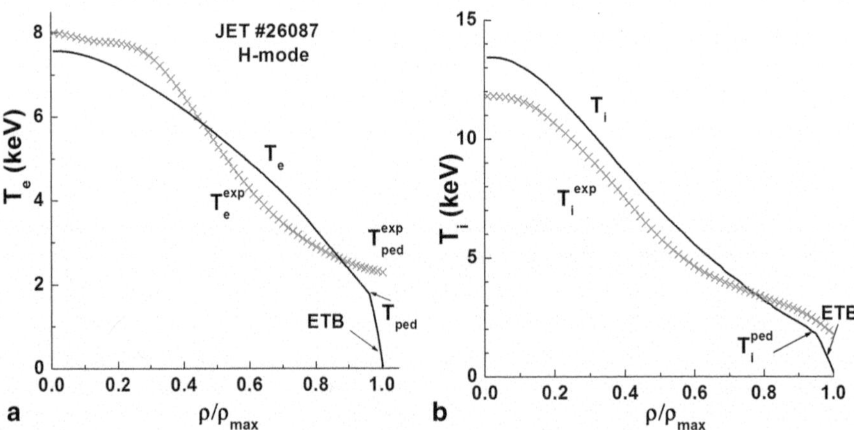

Fig. 6.10 Experimental (*crosses*) and calculated (*solid line*) profiles of the electron (**a**) and ion (**b**) temperatures at the time t=21 s for the JET discharge #26087 with the H-mode

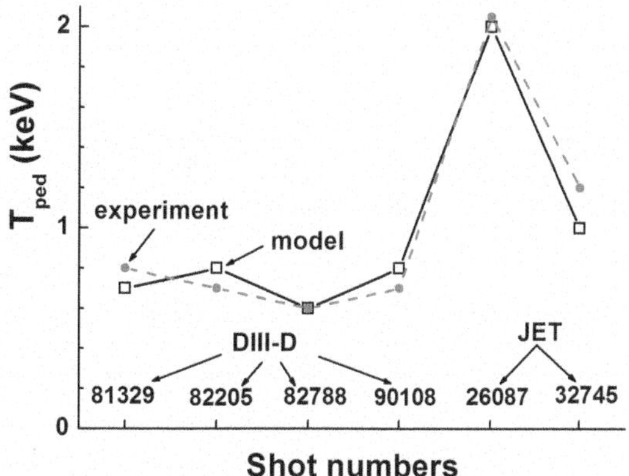

Fig. 6.11 Calculated (*solid line*) and experimental (*dotted line*) values of the temperature pedestal in the steady state for four DIII-D and two JET discharges

Figure 6.11 shows the comparison of the steady state calculated and experimental temperature pedestals for four DIII-D and two JET discharges with very different parameters. It can be seen that the difference does not exceed 10 %.

Figure 6.12a and b show the simulation results for JET discharge #52022 with a non-monotonic power deposition profile (this is also called "off-axis" heating). The power deposition profile to electrons is shown in Fig. 6.12a. In this case, the electron temperature profile in the central region of the plasma $\rho/\rho_{max}<0.6$ is fairly flat, so that the inequality $R_0/L_{Te}<R_0/L_{Tc}$ is satisfied. As a result, the Heaviside func-

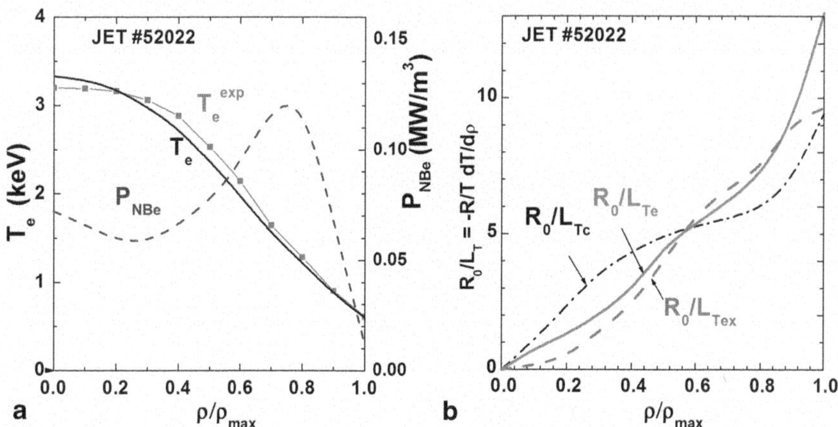

Fig. 6.12 **a** Calculated (*solid line*) and experimental (*line with squares*) electron temperature profiles in the JET discharge #52022 with off-axis heating. Also shown the non monotonic profile of the power deposited into electrons. **b** The relative experimental, R_0/L_{Tex}, and calculated, R_0/L_{Te}, gradients of the electron temperature profile and of the canonical temperature profile R_0/L_{Tc}. At $\rho/\rho_{max}<0.6$ the temperature profile is flat, so $R_0/L_{Te}<R_0/L_{Tc}$ and the Heaviside function in the heat flux (6.4) equals to zero. As a result, the heat transport in this area is low because it is determined by the small second term in Eq. (6.4)

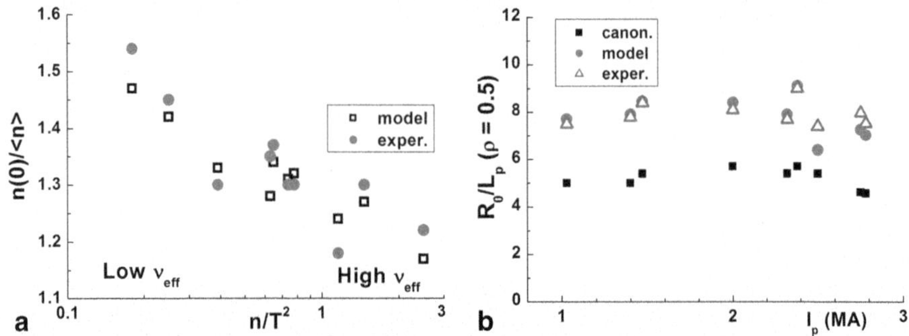

Fig. 6.13 **a** Experimental and calculated peaking parameter of the plasma density profile (the ratio $n(0)/<n>$) for ten JET discharges from Table 6.1 with plasma current $I > 1.5$ MA, versus parameter collisionality n/T^2. Here $<n>$ is the volume averaged plasma density, circles relate to the experiment, squares to the calculation. **b** The relative pressure gradients R_0/L_p at the point $\rho = 0.5$ for nine JET discharges from Table 6.1 depending on the plasma current I. Here, the squares relate to the canonical profiles, triangles—to the experimental profiles and circles—to the calculated pressure profiles

tion in the expression for the heat flux (6.4) is zero in this region and the heat flux is determined by the small second term in this formula.

Figure 6.12b shows the profiles of the dimensionless relative gradients of the experimental electron temperature R_0/L_{Tex}, calculated electron temperature R_0/L_{Te} and the canonical temperature profile R_0/L_{Tc}. In the region $\rho/\rho_{max} < 0.6$ the magnitude of the canonical gradient R_0/L_{Tc} exceeds other gradients. In the region $\rho/\rho_{max} > 0.6$ the inequality $R_0/L_{Te} > R_0/L_{Tc}$ holds, the first term in (6.4) is not zero and the effective heat diffusivity as a result turns out to be large.

Let us now see what the CPTM model learns us about the dependence of the pressure and plasma density profiles on the global plasma parameters. In Fig. 6.13 we show as a function of the collisionality parameter n/T^2 the experimental and calculated values for the ratio $n(0)/<n>$ for 10 JET discharges from Table 6.1 with plasma current $I > 1.5$ MA. Here $<n>$ is the volume-averaged value of the plasma density. The ratio $n(0)/<n>$ is often used to evaluate the peakedness of the plasma density profile. It is very clear that when the value of n/T^2 changes by a factor of 10, the calculated and experimental points remain close to each other, i.e., the model correctly reproduces the reduction of the peakedness of the density profile with increasing values for the collisionality parameter. Figure 6.13b compares the calculated and experimental characteristics of the pressure profile at mid plasma radius. Here we show for 10 discharges from the same Table 6.1 the experimental, calculated and canonical values of the relative pressure gradients R_0/L_p at mid radius, $\rho = 0.5\rho_{max}$. The experimental and calculated values are close to each other and differ from the canonical values by about two units in R_0/L_p. We have already noted earlier that the magnitude of the difference $R_0/L_p - R_0/L_{pc}$ depends on the stiffness of the pressure profile and the auxiliary heating power (see Chap. 5, Sect. 5.5).

So far we have shown examples of ELMy H-mode plasmas. We now turn to the modeling of the ELM free H-Mode discharges and especially JET discharges from

1997 showing record values for the plasma pressure. To obtain ELM free plasmas, a special non-stationary discharge scenario was used that incorporates (1) an increase of the plasma density from $n \sim 2 \times 10^{19}$ m^{-3} to 5×10^{19} m^{-3} during 2–3 s and (2) a simultaneous decrease in the plasma current by 10%. The result is an ELM free H-mode lasting for 1.5–2.5 s.

To compare with experiment, we used in our simulation the primary experimental data, i.e not processed by TRANSP. Therefore, only the electron and ion temperature profile and the total neutral beam power are known. But the ion and electron heat deposition profiles are unknown and we have to use an approximate model for these profiles. In formulating the problem, we introduce the following assumptions:

1. the fraction of lost neutral beam power is 20% of the total power. This is confirmed by detailed calculations for ELMy discharges;
2. the beam power deposition profile $P(\rho)$ (the density of a cloud of hot ions) has a Gaussian shape in radial variable:

$$P(\rho) = P_0 \exp\left(-\frac{\rho^2}{2\rho_0^2}\right), \tag{6.52}$$

where ρ_0 is the width of the profile and P_0 is determined from estimating the total absorbed power;

3. the power depostion profiles from the cloud of hot ions to electrons and ions, P_{NBe} and P_{NBi}, are given by [19]

$$P_{NBi} = \eta_i(\rho)P(\rho), \quad P_{NBe} = \eta_e(\rho)P(\rho), \tag{6.53}$$

with

$$\eta_i = \lambda\left\{\frac{\pi}{3\sqrt{3}} + \frac{1}{3}\ln\left(\frac{1-\sqrt{\lambda}+\lambda}{(1+\sqrt{\lambda})^2}\right) + \frac{2}{\sqrt{3}}\operatorname{arctg}\left(\frac{2-\sqrt{\lambda}}{\sqrt{3\lambda}}\right)\right\},$$
$$\eta_e = 1 - \eta_i. \tag{6.54}$$

For hydrogen $\lambda = \lambda(\rho) = 14.8\, T_e(\rho)/E_0$, and for deuterium $\lambda(\rho) = 18.6\, T_e(\rho)/E_0$, where E_0 is the energy of the main component of the neutral beam.

Figure 6.14a shows the profiles (6.53) for JET discharge #42623 with absorbed power $P_{abs} = P_{NBe} + P_{NBi}$ equal to 16.5 MW. For the same total absorbed beam power P_{abs} the local value of P_{NBi} at $\rho_0 = 0.4$ is twice as large as for $\rho_0 = 0.6$.

In ELM-free discharges the temperature pedestals are quasi-stationary, and they are 3–4 times higher than the averaged pedestal values in ELMy discharges. Therefore, the parameters χ_e^0 and χ_i^0 describing the heat transport inside the ETB with and without ELMs, have to be quite different. Figure 6.14b shows the dependence of the electron and ion temperature pedestals for JET discharge #42623 on χ_e^0 and χ_i^0 as calculated by the model. It is clear from the figure that in order to describe the

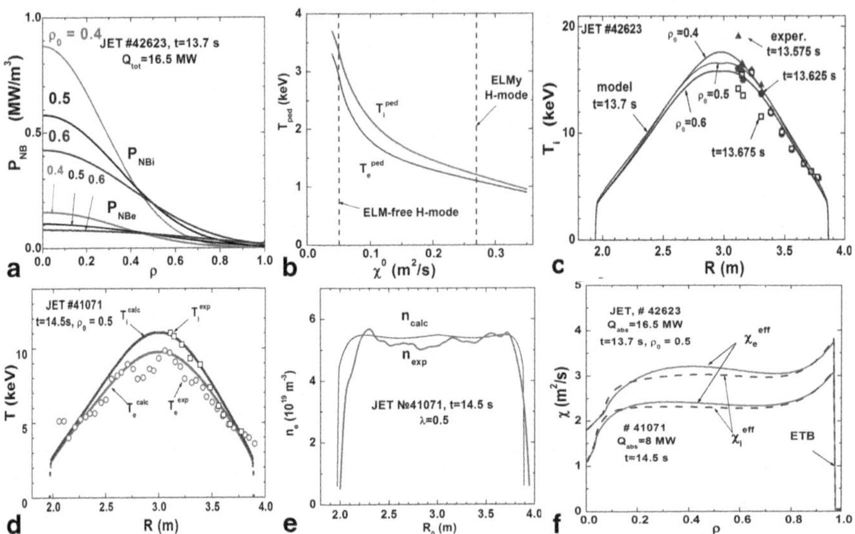

Fig. 6.14 **a** Profiles of the NBI power transmitted to electrons P_{NBe} and ions P_{NBi}, at different peaking of the deposited power profile (ρ_0=0.4, 0.5 and 0.6) for JET discharge #42623. NBI means the Neutral Beam Injection; **b** The calculated dependences of the electron and ion temperature pedestals T_e^{ped} and T_i^{ped} on the values of χ_e^0 and χ_i^0 for the parameters of JET discharge #42623; **c** The calculated ion temperature (*solid curves*) for the JET discharge #42623 at the time instant t=13.7 s at different values of ρ_0=0.4, 0.5 and 0.6. Symbols are the experimental temperatures: *triangles* correspond to t=13.575 s, *circles* to t=13.626 s, *squares* to t=13.675 s. **d** The calculated values of the temperature of ions T_i^{calc} and of electrons T_e^{calc} (*solid lines*) for the JET discharge #41071 at t=14.5 s and ρ_0=0.5. *Squares* are the experimental temperature values of ions T_i^{exp} and *circles* are of the electrons T_e^{exp}; **e** The experimental n_{exp} and calculated n_{calc} profiles of plasma density at t=14.5 s; **f** The profiles of the effective heat diffusivity of electrons χ_e^{eff} (*solid lines*) and ions (*dotted lines*) for the JET discharges #42623 (*upper pair of curves*) and #41071 (*lower pair of curves*). It can be seen that within the ETB the heat diffusivity is very low of order 0.04 m²/s

temperature profiles of ELM-free discharges the parameters χ_e^0 and χ_i^0 should be 8–10 times smaller as compared to ELMy discharges as given by expression (6.37). The density pedestals in ELM-free discharges are about the same as in ELMy discharges, so the values of D^0 remain those given by formula (6.38).

A comparison between the calculated profiles for the ion temperature in JET discharge #42623 with $\chi_e^0 = \chi_i^0 = 0.04(\mathrm{m}^2/\mathrm{s})$ and the experimental ones is shown in Fig. 6.14c. The experimental points are taken at three successive time points (as indicated in the figure), in a total time interval of 100 ms. The results of the calculation for ρ_0=0.4, 0.5 and 0.6 are shown. All the calculated curves coincide reasonably well with the experimental ones. As a particular conclusion, it is clear that the simulation results using a stiff set of equations are weakly dependent on the power deposition profile for a centrally peaked deposition profile. For off-axis heating the results may be different.

Figure 6.14d shows the experimental and calculated electron and ion temperature profiles for JET discharge #41071 (with P_{abs}=8 MW) for ρ_0=0.5 and the same

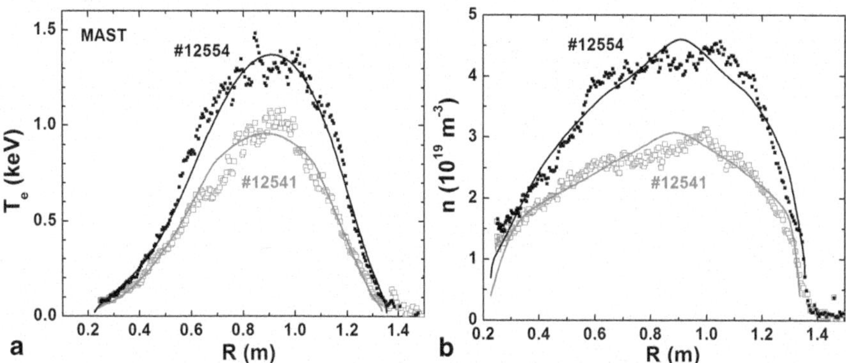

Fig. 6.15 Calculated (*solid lines*) and experimental (*squares*) profiles of the electron temperature (**a**) and plasma density (**b**) in the MAST discharges #12554 (*black symbols and lines*) and #12541 (*open symbols and lines*) in the L-mode

values of heat diffusivities $\chi_e^0 = \chi_i^0 = 0.04(\text{m}^2/\text{s})$. The comparison of the calculated and experimental profiles of the plasma density for this discharge is provided in Fig. 6.14e Again we obtain good agreement between the experimental and calculated profiles.

Note that the agreement between experiment and calculations for temperature and density profiles over the full plasma cross section, including the ETB, is obtained adjusting only one single parameter $(\chi^0 = \chi_e^0 = \chi_i^0)$ in the model. The profiles for the electron and ion effective heat diffusivity for the ELM-free discharges of Fig. 6.14d and e are shown in Fig. 6.14f. It is clear from the figure that when the heating power is doubled (the transition from #41071 to #42623), the effective heat diffusivity only increases by a factor $\sqrt{2}$. Note however that the model for the ELM-free H-mode has no realistic "predictive" quality because the database used to set up the model is rather limited.

Let's now turn to the simulation of discharges on MAST, a device with low aspect ratio $A = a/R_0 \sim 1.4$ and low toroidal magnetic field $B_t \sim 0.5$ T. The plasma parameters here are very different from the ones in conventional tokamaks with an aspect ratio $A \sim 3$–5 and toroidal field between 2 and 4 T. First we consider discharges in the L-mode. Figure 6.15 shows the experimental and calculated profiles of the electron temperature (a) and plasma density (b) for discharges #12541 and #12554. The profiles are measured by the Thomson scattering diagnostic with high spatial resolution (300 points along the major radius). Neutral Beam Injection (NBI) was used for plasma heating with a total heating power of about 1.3 MW.

Figures 6.16 and 6.17 show electron temperature and density profiles for H-mode discharges #13035 and #13026. The electron temperature pedestal is low for both discharges. Figure 6.16b shows that the plasma density profile has a very rectangular shape, and is very flat for most of the plasma cross section with a large gradient within the ETB. In discharge #13026 the plasma density was rapidly increased using strong gas puffing. Two processes characterisize the density balance at the plasma edge. First, the large density pedestal leads to charge exchange and

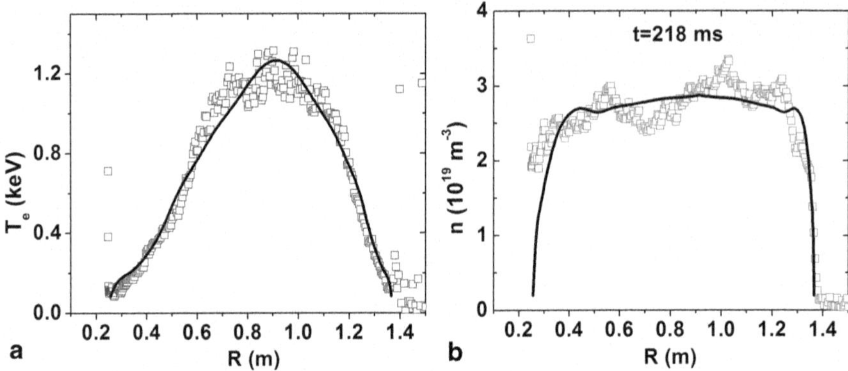

Fig. 6.16 Calculated (*solid line*) and experimental (*squares*) profiles of the electron temperature (**a**) and plasma density (**b**) in the MAST H-mode discharge #13035. It is seen that the density profile is very flat with high pedestal, but the temperature pedestal does not exceed 0.2 keV, so the temperature profile is strongly peaked, due to conservation of the pressure profile

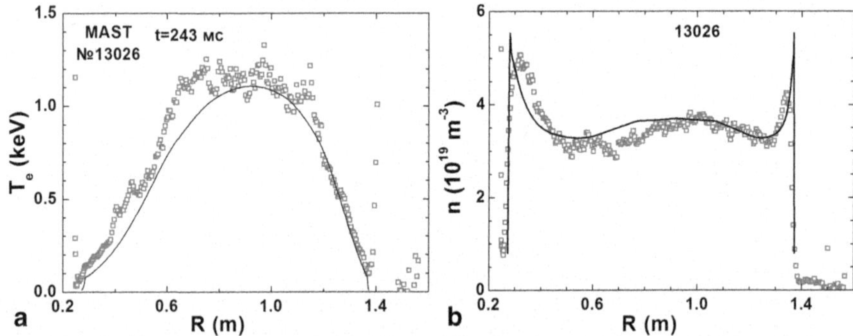

Fig. 6.17 Calculated (*solid line*) and experimental (*squares*) profiles of the electron temperature (**a**) and plasma density (**b**) in the MAST H-mode discharge #13026. At the high ramp up of the averaged density (high gas puff) the "ears" are formed on the density profile

ionization of neutrals from the wall inside and close to the transport barrier. Second, the slow diffusion, small convection and the strong source of particles hinder the charged particles to penetrate quickly into the inner layers of the plasma. As a result, so-called "ears" are formed at the density profile of the plasma, that are visible on the experimental and calculated density profiles in Fig. 6.17b.

Next we have a look at modeling results for internal transport barriers (ITB). At MAST an ITB is often formed for the ion temperature. To simulate the formation of the barrier we must first construct a function $z_{0i}(\rho)$, as given by expression (6.43). In discharge #8575 early NBI heating led to the formation of a non-monotonic profile for the magnetic shear s with a minimum around $\rho/\rho_{max} = 0.35$. The corresponding form for the function $z_{0i}(\rho)$ is shown in Fig. 6.18a. Figure 6.18b shows the experimental and calculated profiles of the ion temperature for the three time points

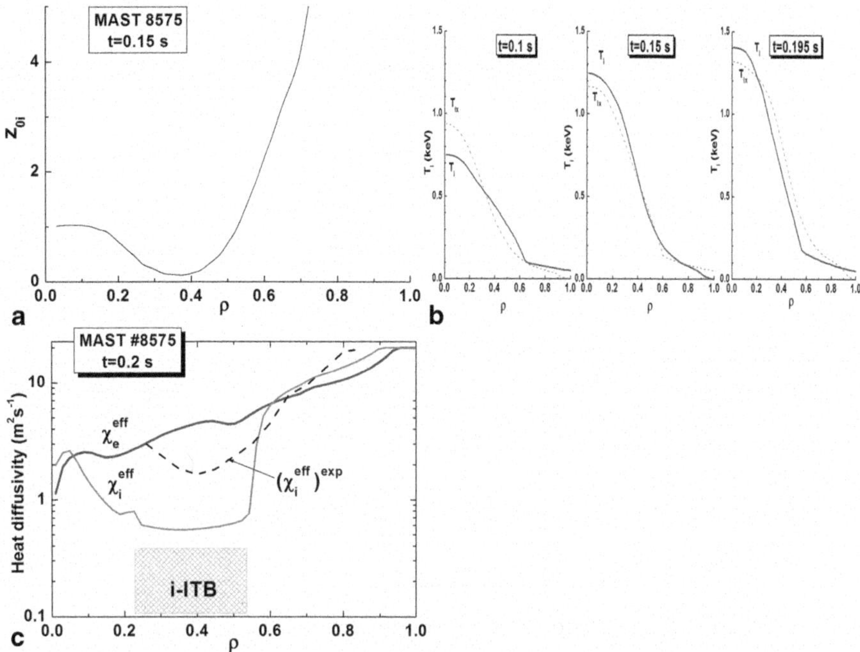

Fig. 6.18 **a** The behaviour of $z_{0i}(\rho)$ at the negative shear $s = \rho/q\ \partial q/\partial\rho$ adopted in the simulation of internal transport barrier (ITB) in the MAST discharge #8575. **b** Experimental T_{ix} and calculated T_i profiles, of the ion temperature at time t=0.1, 0.15 and 0.195 s at the formation of ion ITB. **c** Calculated (*solid curves* for electrons and ions) and experimental (*dotted line* for ions), the effective heat diffusivity coefficients. The barrier is in the range $0.22<\rho/\rho_{max}<0.52$, where the magnetic shear s is negative

(t=0.1, 0.15 and 0.195 s) in the discharge. At $t\sim0.195$ s a steep gradient is formed in the ion temperature profile for $0.22<\rho/\rho_{max}<0.52$ and the value of the dimensionless ratio R_0/L_{Ti} is between 15 and 18.

Profiles for the effective heat diffusivities for this discharge are shown in Fig. 6.18c. Within the ITB the calculated value of χ_i^{eff} is reduced by an order of magnitude compared to its values in the gradient zone $\rho/\rho_{max}\sim0.7$. The discrepancy between the calculated χ_i^{eff} and experimental values $(\chi_i^{eff})^{exp}$, visible in Fig. 6.18c is due to the low precision for the calculated power deposition profiles and measurements of the gradient $\frac{\partial T_i}{\partial\rho}$ inside the ITB.

Figure 6.19 shows an example for the modeling of the internal transport barrier in the ion transport channel (i-ITB) for JET discharge #40847. The calculated and experimental ion and electron temperature profiles (a) and experimental effective heat and particle diffusivity coefficients (b) are shown. For the function z_{0i} expression (6.43) with coefficients from Table 6.2 was used. The ion ITB in the region $0.4<\rho/\rho_{max}<0.55$ is clearly visible. The calculated position is slightly shifted outwards ($\delta\rho\sim0.1a$) compared to the experimental position.

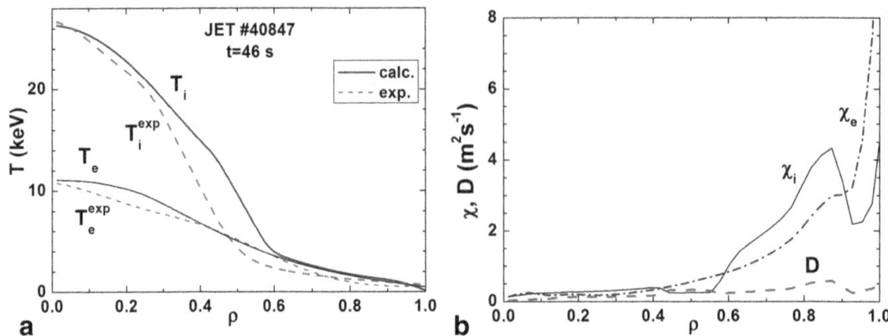

Fig. 6.19 Experimental and calculated temperature profiles of electrons and ions (**a**) and profiles of the effective coefficients of heat and particle diffusivities (**b**), in the JET discharge #40847 at time t=46 s, when the ITB was formed in the ion transport channel in the region $0.45<\rho/\rho_{max}<0.55$

6.8 Remarks on the Stiffness of the Ion Temperature Profile

6.8.1 Radial Dependence of the Stiffness

In Sect. 5.3, where we determined the transport coefficients, we supposed that the stiffness coefficient κ_i^{PC} is constant over the plasma cross section (5.12). Such an assumption dates back from earlier times in fusion research [2], when no reliable measurements of the ion temperature gradient along the plasma radius were available. The strong stiffness of the heat conduction Eq. (5.2) has also contributed to this assumption, leading to a weak dependence of the calculated temperature profiles on the absolute values of κ_i^{PC}. This situation is drastically changed in recent years, and reliable measurements of both the ion temperature profiles and the temperature gradients, are available. A more detailed treatment of this subject is thus possible.

Let us refer again to Fig. 5.1, which shows the dependence of the heat flux q_T on the relative temperature gradient T'/T. The slope of branch (II) is determined by the stiffness of the temperature profile. The higher the value of the coefficient κ_i^{PC}, the steeper is branch (II) for the ions. In [20] we analyzed the ion temperature gradient for a series of JET discharges with different values for the heating power but otherwise similar plasma parameters and found that the slope of branch (II) increased with radius ρ. The difficulty of measuring this slope at the plasma edge did not allow a precise value for κ_i^{PC} over the whole plasma cross section. However it was clear that the stiffness coefficient κ_i^{PC} is several times smaller in the plasma core than in the plasma edge.

How to reconcile the behavior of the stiffness coefficient in the experiment with our model (5.12), which suggests that the stiffness is constant along the radius? After all, we have seen that in many cases, the simulation results are in reasonable agreement with experiment. The answer is as follows. As we have seen, the

Table 6.3 The main parameters of the selected discharges averaged over the time interval t=8–9 s

Dicharge No	n	P_{NBI}	P_{ICRH}	P_{tot}	τ_E	Central rotation frequency ω	Main heating
	10^{19} m^{-3}	MW	MW	MW	s	10^4 rad/s	
50628	5.4	6.16	3.17	9.04	0.41	7.8	NBI
52097	4.55	3.85	9.55	13.1	0.3	2.9	ICRH

stiffness coefficient defined by (5.12), allows a reasonably description of the behavior of the stiffness in the gradient zone of the plasma $0.5<\rho/\rho_{max}<0.7$. In the central zone $\rho/\rho_{max}<0.5$ the experimental stiffness is less, and in the region $\rho/\rho_{max}<0.3$ the sawtooth oscillations distort the temperature profile. As a result, the time-averaged experimental temperature profile is close to the calculated one. It is clear that for plasma regimes without sawtooth oscillations the difference between the calculated results and the experimental ones will be more pronounced in the central zone of the plasma.

In this context, let us also discuss the results of the simulation of two JET discharges with the same minor radius, magnetic field and plasma current: R=2.92 m, a=0.95 m, B_0=2.85 T, I=2.8 MA; differences in other plasma parameters are given in Table 6.3. These discharges were heated with a combination of NBI and ion cyclotron resonance heating (ICRH). In calculations of P_{tot} in Table 6.3 the radiated power was taken into account.

Transport calculations were done with the help of an artificial piece-wise constant "stiffness function" for the ions only:

$$S(\rho,\rho_0)=S_0 \text{ at } 0<\rho<\rho_0,\ S(\rho,\rho_0)=S_1 \text{ at } \rho_0<\rho<\rho_{max} \tag{6.55}$$

and using the product $S\kappa_i^{PC}$ in the ion heat flux (k=i) (5.8) instead of κ_i^{PC}. The expression for the electron heat flux remains unchanged.

First, we set

$$\rho_0=\rho_{max} \tag{6.56}$$

and simulate discharge #50628 for different values of S_0 (the value of S_1 is absent in the model (6.56)). Figure 6.20 shows the ion temperature profiles, obtained using the non-linear model for the heat fluxes (6.4), assuming (6.55) and (6.56) for S_0=0.2, 1, 2, 5 and 10. The figure also shows the experimental ion temperature profile. The best approximation to the experimental profile in the gradient zone is obtained for S_0=1.

The profiles of the normalized relative gradients of the ion temperature $R_0/L_{Ti}=-R_0T_i'/T_i$ at S_0=0.2, 1, 2, 5 and 10 assuming (6.55−6.56) are shown in Fig. 6.21. The behavior of normalized relative gradients for the canonical temperature profile $R_0/L_{Tc}=-R_0\ T_c'/T_c$ is also shown. It is clear from the figure that by increasing the stiffness parameter S_0 the calculated gradient profiles approach the

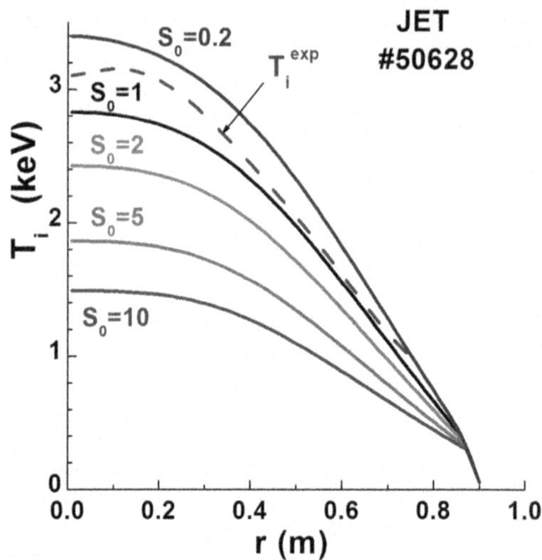

Fig. 6.20 The profiles of ion temperature for JET discharge #50628 in the model (6.55–6.56) at different values of the stiffness parameter S_0. The experimental ion temperature profile is shown also by *dotted line*

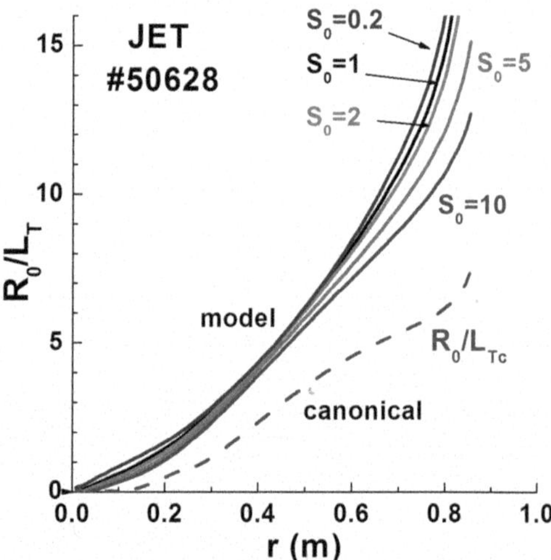

Fig. 6.21 The relative ion temperature gradients R_0/L_{Ti} for JET discharge #50628 at different S_0 calculated in the model (6.55–6.56). The correspondent curve for the canonical temperature profile R_0/L_{Tc} is also shown

canonical ones. In this particular case, however, the calculated value of the ion temperature at the pedestal, shown in Fig. 6.20, turns out to be low, so the distance between the calculated R_0/LT_i and canonical R_0/LT_c curves in Fig. 6.21 is significant, even at high S_0.

The ion temperature profiles obtained assuming

$$\rho_0 = 0.5\rho_{max} \tag{6.57}$$

and for different values of S_0 and $S_1 = 1$ are shown in Fig. 6.22. Figure 6.23 shows the normalized relative gradients of the ion temperature at different values of stiff-

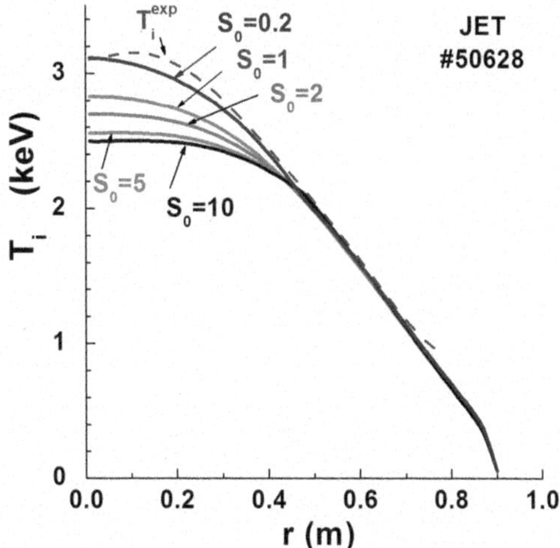

Fig. 6.22 The same as in Fig. 6.20, but for the model (6.55), (6.57). It is seen that the temperature changes in the region $\rho/\rho_{max}<0.5$ only

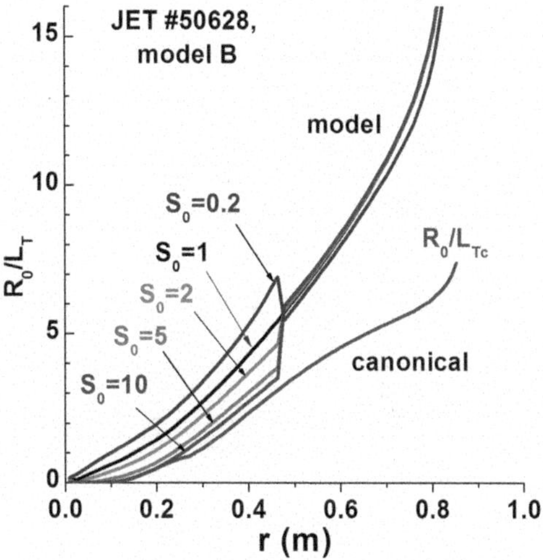

Fig. 6.23 The same as in Fig. 6.21, but for the model (6.57)

ness parameter S_0 at fixed $S_1=1$. It is seen that the assumption (6.57) influences the shape of the profile almost exclusively in the core plasma ($\rho/\rho_{max}<0.5$).

Now we compare the results obtained by the models (6.56) and (6.57). The dependencies of the ratios $T_{i0}(S_0)/T_{i0}(S_0=1)$ and $\tau_E(S_0)/\tau_E(S_0=1)$ on the parameter S_0 for both models are shown in Figs. 6.24 and 6.25. It is clear from the figures that increasing the stiffness function S by a factor of 10 over the whole plasma cross section (model (6.56)) reduces the central ion temperature $T_i(0)$ by 50 %. A similar increase of S_0 by a factor of 10 in the core plasma (model (6.57)) reduces $T_i(0)$ only by 15 %. The variation in the energy confinement time is even smaller. Using expression (6.56) in the model under otherwise similar conditions, τ_E is changed by 40 %; however, using expression (6.57) for the modelling the variation is limited to 2–3 %.

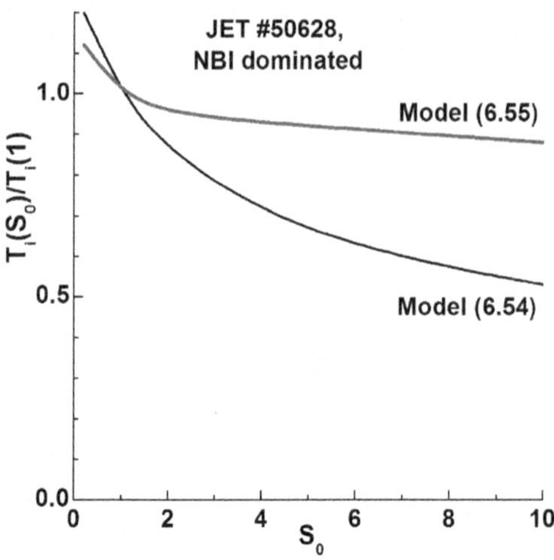

Fig. 6.24 The relative central ion temperature $T_i(S_0)/T_i(1)$ in the models (6.56) and (6.57) at different values of stiffness parameter S_0 for JET discharge #50628

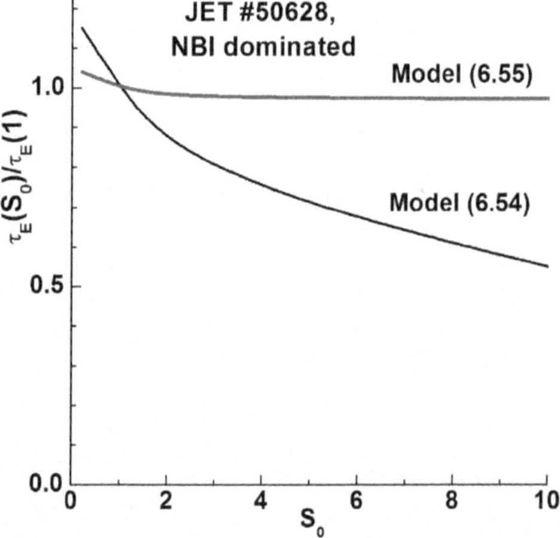

Fig. 6.25 The relative value of the energy confinement time $\tau_E(S_0)/\tau_E(1)$ in the models (6.56) and (6.57) at different values of stiffness parameter S_0 for JET discharge #50628

6.8.2 Dependence of the Stiffness on the Toroidal Rotation Velocity

So far we have discussed profile stiffness in discharges with predominantly NBI heating. In such discharges the fast beam particles transfer a high torque on the plasma and induce a fairly high toroidal plasma rotation. In JET with NBI powers of about 10 MW the velocity of toroidal rotation reaches values $v_t \sim 200$–300 km/s. To estimate the influence of the toroidal plasma rotation on energy confinement a dedicated set of experiments were performed at JET with a decrease in the velocity of rotation while keeping the total heating power constant. This was achieved by replacing part of the NBI power by ion-cyclotron resonance heating (ICRH).

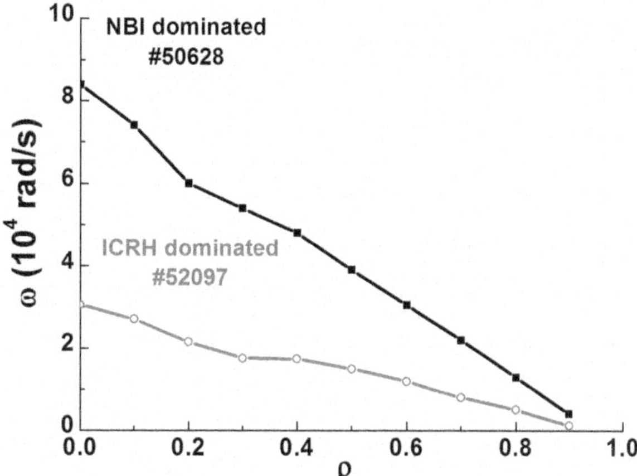

Fig. 6.26 The angular velocity of plasma toroidal rotation for the JET discharge #50628 with predominantly NBI heating, and for discharge #52097 with predominantly ICRH heating

The difference with NBI is that ICRH heating has only a minimal influence on torque and rotation in the plasma. Applying this method of mixed heating allows one to change the rotation velocity by a factor 3–5. The experiment shows that, despite the large decrease in the rotation velocity, the energy confinement time τ_E only decreases by 10% [21]. According to Fig. 6.25, to obtain a similar decrease of τ_E the stiffness coefficient κ_i^{PC} needs only to increase by a factor 1.5–2 in the gradient zone.

The discharges with different rotation velocities described above were used to analyze the profile stiffness of the ion temperature at different toroidal plasma rotation velocities [20, 22, 23]. Summarized, the decrease in rotation velocity leads to a strong increase of the ion temperature stiffness κ_i^{PC} in that part of the plasma cross section where the magnetic shear $s = \rho/q\ \partial q/\partial\rho$ is sufficiently low. Experimentally no change was observed for the stiffness in the gradient zone. This is possibly due to the large experimental errors in the measurement of stiffness at the plasma edge.

To estimate the value of stiffness change at the plasma edge with variation of rotation velocity we proceed as follows. We compare JET discharge #50628 with predominantly NBI heating and high rotation velocity with JET discharge #52097 with predominantly ICRH heating and low rotation velocity (Table 6.3). Toroidal rotation velocity profiles for these discharges are shown in Fig. 6.26.

Simulation results for discharge #52097 with low rotation velocity are shown in Fig. 6.27 (using expression (6.56) in our model), and Fig. 6.28 (using expression (6.57) for modeling). Figure 6.27, shows clearly that for $S_0=1$ the calculated curve is slightly above the experimental one, with the largest difference in the gradient zone. In the core plasma the calculated curve is close to the experimental one. Assuming $S_0=2$, the situation is reversed: in the gradient zone the calculated and experimental curves are very similar, but the central temperatures are markedly different.

Figure 6.28 shows that in the gradient zone, where $S_1=1$, there is a large difference for all shown curves between the calculations and experiment. This analysis thus shows that for a correct description of the experiment with low rotation velocity the stiffness should be increased in the gradient zone by a factor 1.5–2 and

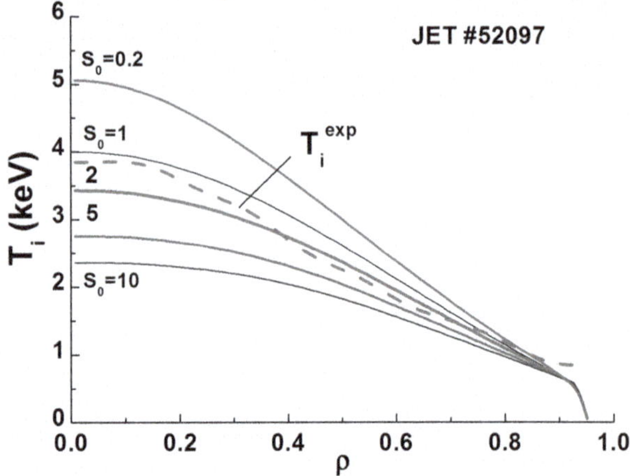

Fig. 6.27 Ion temperature profiles for the JET discharge #52097 in the model (6.56) at the different values of stiffness parameter S_0

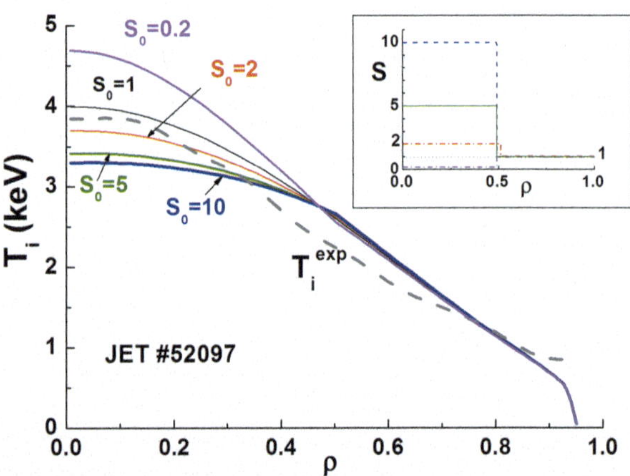

Fig. 6.28 The profiles of ion temperature are shown for the JET discharge #52097 in the model (6.57) at the different values of stiffness parameter S_0. Corresponding profiles of the stiffness function S are shown in the inset

together with a reduction in the stiffness in the core plasma by a factor 1.5–2. Thus, the stiffness in core plasma is 3–4 times smaller than in the edge plasma. Thus both effects, the decrease of τ_E by 10 % and an improved description of the ion temperature profile$_i$, are consistent with a doubling of the stiffness in the gradient zone when the rotation velocity decreases by a factor of three (Fig. 6.26),

Figure 6.29 shows a comparison of the calculated and experimental ion temperature profiles obtained using different assumptions for the stiffness as discussed above in our model. For the simulation of JET discharge #50628 with predominantly NBI heating expression (6.57) is assumed with $S_0=0.3$, $S_1=1$, and for the simulation of discharge #52097 with predominantly ICRH heating the same model is used, but with $S_0=0.6$ and $S_1=2$. There is a fair agreement between the calculated curves and the experimental ones.

To summarize this Section, we arrive at the following conclusions:

- The stiffness in the gradient zone is at least 3–4 times larger than the stiffness in the core plasma;

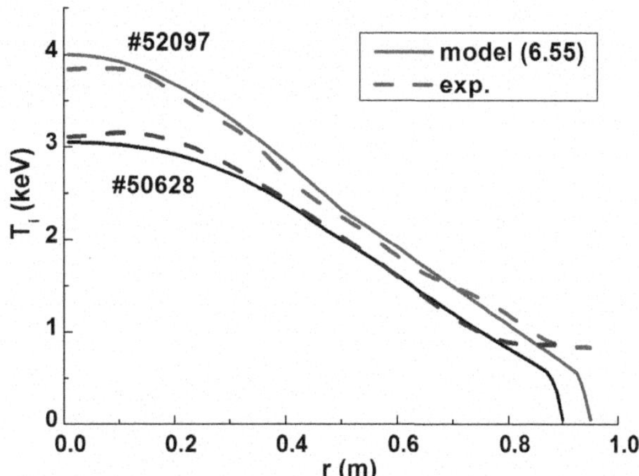

Fig. 6.29 The ion temperature profiles. Solid lines are the model calculations (6.57) with $S_0=0.6$, $S_1=2$ for the JET discharge #52097 with a low rotation velocity and $S_0=0.3$, $S_1=1$ for discharge #50628 with high rotation velocity. Dotted lines are the experiment for these discharges

- The stiffness is reduced by a factor of 2 over the whole plasma cross section when the rotation velocity is increased by a factor of 3.

Of course, based on only two discharges one can only determine trends in the change of the stiffness of the ion temperature profile. To determine more detailed information on the behaviour of the stiffness coefficient κ_i^{PC} and its profile when the toroidal rotation velocity is changed, will require a large database of discharges from several devices. However, the results described above give an order of magnitude of the changes that one can expect.

Experiments carried out recently in DIII-D to study the profile stiffness of the ion temperature are described in reference [24]. The main result from this report is a confirmation that the stiffness in the gradient zone ($\rho \sim 0.7$) is significantly higher (by a factor 8–10) than in the plasma core ($\rho \leq 0.4$). In the central part of plasma the ion temperature profile is "soft", as there are large variations in the temperature gradient with relatively small changes in the heating power. The reduction of the toroidal rotation velocity in DIII-D was achieved by applying simultaneously various amounts of co- and counter- neutral beams. This conclusion was confirmed by a remarkable increase in the ion temperature profile stiffness when decreasing the toroidal rotation velocity.

6.9 Approximation for the Pedestal Values Based on the Experimental Data

6.9.1 General Expressions

Let us consider once more the linear version of the CPTM (see Chap. 5). This means that we exclude the external transport barrier from the considerations below, but will try to find the boundary conditions close to position of the pedestals. For the heat flux we use expression (5.8). During on-axis heating the argument of Heaviside function is positive, so $H(x)=1$. The second term in (5.8) can be omitted due to inequality $\kappa_k^{PC} \gg \kappa_k^0$. We omit also the third term in (5.8) due to (5.16−5.17). As a result, the temperature profile is monotonic. Supposing $T_c(0)=T_k(0)$, we obtain

$$\frac{T_c'}{T_c}-\frac{T_k'}{T_k}=\frac{\partial}{\partial\rho}\ln\left(\frac{T_c(\rho)}{T_k(\rho)}\right)=\frac{\partial}{\partial\rho}\ln\left(\frac{T_c(\rho)}{T_c(0)}\cdot\frac{T_k(0)}{T_k(\rho)}\right)\geq 0 \tag{6.58}$$

Let us introduce the notation

$$f(\rho)=\frac{T_c(\rho)/T_c(0)}{T_k(\rho)/T_k(0)} \tag{6.59}$$

Due to (6.58−6.59), we have

$$\frac{\partial}{\partial\rho}\ln f=\frac{1}{f}\frac{\partial f}{\partial\rho}\geq 0, \quad f(\rho)\geq 0, \quad f(0)=1, \tag{6.60}$$

Therefore, $df/d\rho \geq 0$. Hence

$$f(\rho)\geq 1. \tag{6.61}$$

Defining the *normalized temperature* $\phi(\rho)=1/f(\rho)$, we obtain

$$1\geq\phi(\rho)=\frac{T_k(\rho)/T_k(0)}{T_c(\rho)/T_c(0)}\geq 0 \tag{6.62}$$

It is evident that $d\phi/d\rho \leq 0$. To estimate the absolute value of the pedestal $T_k(\rho_{max})$ we need to find $\phi(\rho_{max})$. To this end we analyse the behavior of $\phi(\rho)$, using for $T_k(\rho)$ the experimental temperature profile.

6.9.2 Estimates for the Normalized Temperature Based on JET Experimental Data

Table 6.1 shows the main parameters of JET discharges used for the analysis [13]. Since the experimental data for electron and ion temperatures at the plasma edge have usually large errors, we use in what follows, the normalized sum of the electron and ion temperatures instead of $\phi(\rho)$ to improve the reliability:

$$T_n(\rho)=\frac{T_e(\rho)+T_i(\rho)}{T_e(0)+T_i(0)}\cdot\frac{T_c(0)}{T_c(\rho)} \tag{6.63}$$

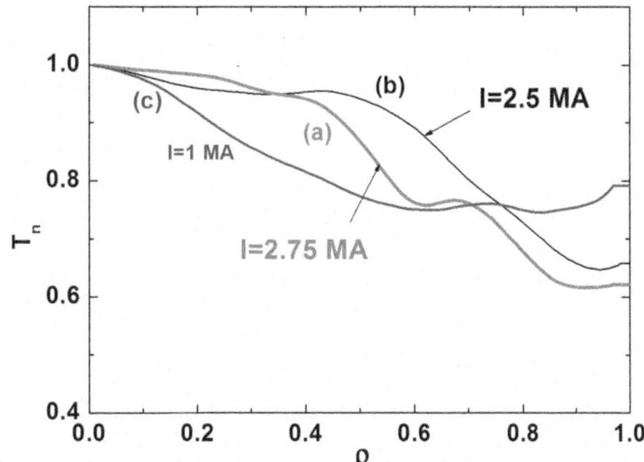

Fig. 6.30 The experimental profiles of the sum of the electron and ion temperatures normalized to the canonical temperature profile

$T_n(\rho) = \frac{T_e(\rho)+T_i(\rho)}{T_e(0)+T_i(0)} \cdot \frac{T_c(0)}{T_c(\rho)}$

for three JET discharges #61103 (**a**), #61138 (**b**) and #61526 (**c**) at the time moments $t = 15$, 20 and 22 s correspondingly

and we make no distinction between the electron and ion temperature pedestals. For T_e and T_i we will use the measured values $T_e^{exp}(\rho)$ and $T_i^{exp}(\rho)$, processed by TRANSP. In Table 6.1 the value of P_{NB} is total NBI power deposited into plasma, I is the plasma current, $\bar{n}$ is the chord-averaged plasma density, $T_n(\rho_{max})$ is the function (6.63) at the plasma edge averaged over a time interval of about 1 s.

Figure 6.30 shows the profiles of the function $T_n(\rho)$ for JET discharges #61103 at high plasma current $I = 2.75$ MA and high plasma density $\bar{n} = 6.4 \times 10^{19}\,\mathrm{m}^{-3}$, #61138 at a plasma current $I = 2.5$ MA and very high density $\bar{n} = 9.9 \times 10^{19}\,\mathrm{m}^{-3}$ and #61526 at low plasma current $I = 1$ MA and low density $\bar{n} = 2.7 \times 10^{19}\,\mathrm{m}^{-3}$.

The time evolution of $T_n(\rho, t)$ at the two radial positions $\rho = \rho_{max}$ and $\rho = 0.8\rho_{max}$ for the same discharges is shown in Fig. 6.31. These functions differ only by 5–10% from a constant function.

Figure 6.32 shows the time averaged values of

$$T_n^{av}(\rho_{max}) = \langle T_n(\rho_{max}, t) \rangle \tag{6.64}$$

for 13 discharges with descending plasma current values shown in Table 6.1. The averaging is provided over time intervals of about 1 s around the time points shown in Table 6.1. For some discharges in this figure also the chord averaged values of the density, the values of the neutral beam heating power P_{NB} and the values of safety parameter at the edge q_a are shown. It is clear from this figure that the values of T_n slightly decrease with increasing current and strongly decrease with decreasing density.

Figure 6.33 shows the values of $T_n^{av}(\rho_{max})$ as a function of the plasma density for the discharges from Table 6.1 at high current $I > 2$ MA. The figure shows clearly that the pedestals increase monotonically with increasing density keeping the value of T_n^{av} in the range $0.4 < T_n^{av}(\rho_{max}) < 0.6$.

To estimate the plasma density pedestal we need to analyze the experimental behavior of the normalized pressure p_n which is similar to the normalized temperature T_n

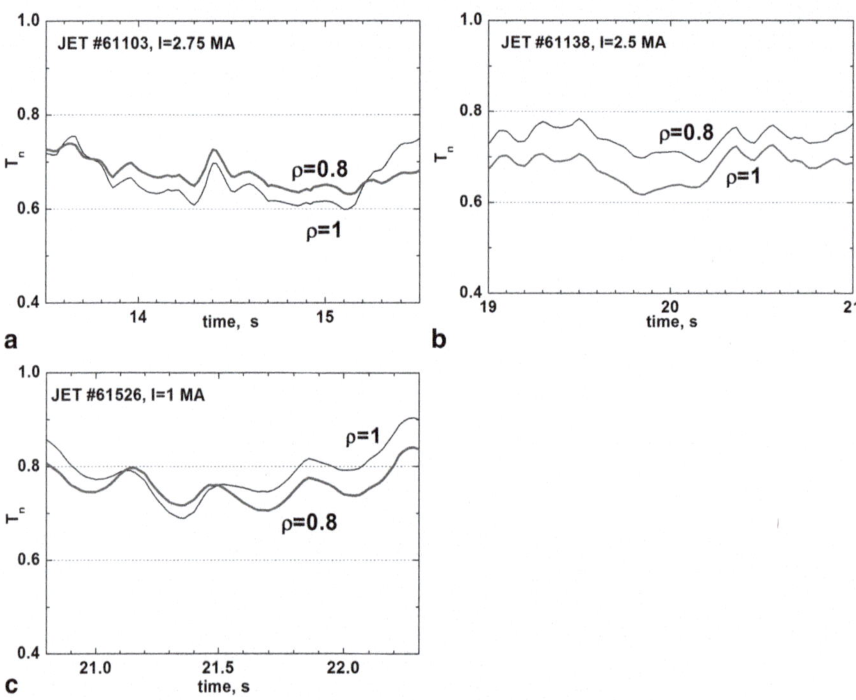

Fig. 6.31 The time behavior of the normalized experimental temperature T_n at the normalized spatial points $\rho=0.8$ and $\rho=1$ (pedestal) for JET discharges #61103 (**a**), #61138 (**b**) and #61526 (**c**)

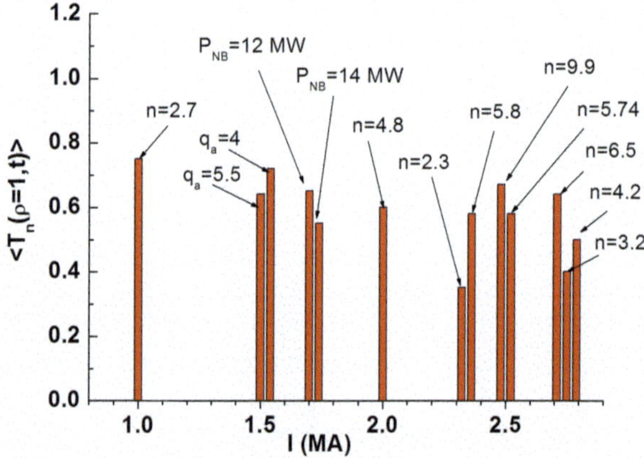

Fig. 6.32 The normalized, averaged over time temperature pedestal T_n for JET discharges from Table 6.1 with different plasma current. The values of chord-averaged densities, deposited power and q_a are also shown for some shots

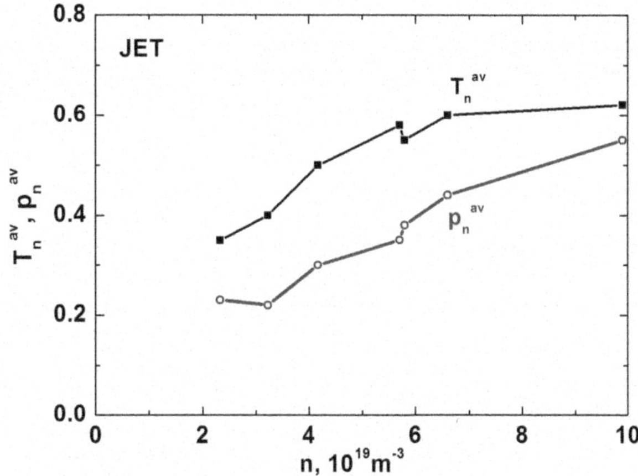

Fig. 6.33 The normalized, averaged over time temperature T_n and pressure p_n pedestals for JET discharges from Table 6.1 with high plasma current I>2.2 MA versus plasma density

$$p_n(\rho_{max}) = \frac{p(\rho_{max},t)/p(0)}{p_c(\rho_{max})/p_c(0)} \tag{6.64}$$

The dependence of $p_n^{av}(\rho_{max})$ on the plasma density for plasma currents I>2 MA is also shown in Fig. 6.33. The curves shown in this figure are not only suitable for JET but also for other devices since the normalized values of T_n and p_n are independent of the plasma geometry and heating power as these values are taken into account in the calculations of $T_{e,i}(0)$, $T_c(0)$, $p(0)$ and $p_c(0)$. The curves from Fig. 6.33 can be used in transport calculations to estimate the temperature and density pedestals at sufficiently high currents according to $I/a^2>2$ (I in MA, a in m).

References

1. Wagner, F., et al.: Regime of improved confinement and high beta in neutral-beam-heated divertor discharges of the ASDEX Tokamak. Phys. Rev. Lett. **49**, 1408 (1982)
2. Dnestrovskij, Yu.N., Lysenko, S.E., Tarasyan, K.N.: Improved confinement regimes within the transport model of canonical profiles. Nucl. Fusion **35**, 1047 (1995)
3. ITER Physics Basis, Chapter 2, Plasma confinement and transport . Nucl. Fusion **39**, 2175 (1999)
4. Dnestrovskij, Yu.N., Dnestrovskij, A.Yu., Lysenko, S.E.: Self-organization of plasma in a tokamak. Plasma Phys. Rep. **31**, 529 (2005)
5. Dnestrovskij, Yu.N., et al.: Analysis of pressure profiles and transport simulations of MAST discharges. Plasma Phys. Control. Fusion **49**, 1477 (2007)
6. Dnestrovskij, Yu.N., et al.: Canonical profiles and transport model for the toroidal rotation in tokamaks. Plasma Phys. Control. Fusion **53**, 085025 (2011)
7. Progress in ITER Physics Basis. Nucl. Fusion **47**, S109 (2007)
8. Kirk, A., et al.: A comparison of H-mode pedestal characteristics in MAST as a function of magnetic configuration and ELM type. Plasma Phys. Control. Fusion **51**, 065016 (2009)
9. Hughes, J.V., et al.: Power requirements for superior H-mode confinement on Alcator C-Mod experiments in support of ITER. Nucl. Fusion **51**, 083007 (2011)

10. Connor, J.W., Hastie, R.J., Wilson, H.R.: Magnetohydrodynamic stability of tokamak edge plasmas. Phys. Plasma. **5**, 2687 (1998)
11. Groebner, R.J., et al.: Progress towards a predictive model for pedestal height in DIII-D. Nucl. Fusion **49**, 085037 (2009)
12. Groebner, R.J., Osborne, T.H., Leonard, A.W., Fenstermacher, M.E.: Temporal evolution of H-mode pedestal in DIII-D. Nucl. Fusion **49**, 045013 (2009)
13. Dnestrovskij, Yu.N., et al.: Application of canonical profiles transport model to the H-mode shots in Tokamaks. Plasma Phys. Rep. **36**, 645 (2010)
14. Dnestrovskij, Yu.N., et al.: Simulation of internal transport barriers with the help of the transport model of canonical profiles. Plasma Phys. Rep. **32**, 1 (2006)
15. Lebedev, S.V., Andrejko. M.V., Ashkinazi, L.G., Golant, V.E., et al.: H-mode studies in Tuman-3 and Tuman-3M. Plasma Phys. Control. Fusion **38**, 1103 (1996)
16. Alikaev, V.V., et al.: Investigation of the H-mode during ECRH in the T-10 tokamak. Plasma Phys. Rep. **26**, 917 (2000)
17. G. Tresset et al.: A dimensionless criterion for characterizing internal transport barriers in JET. Nucl. Fusion **28**, 520 (2002)
18. The ITER 1D Modelling Working Group: Boucher, D., Connor, J.W., Houlberg, W.A., et al.: The international multi-tokamak profile database. Nucl. Fusion **40**, 1955 (2000). http://tokamak-profiledb.ukaea.org.uk/
19. Dnestrovskij, Yu.N., Kostomarov, D.P.: Mathematical modeling of plasma. Nauka, Moscow (1993) in Russian
20. Mantica, P., et al.: Experimental study of the ion critical-gradient length and stiffness level and the impact of rotation in the JET tokamak. Phys. Rev. Lett. **102**, 175002 (2009)
21. Versloot, T.W., et al.: Comparison between dominant NB and dominant IC heated ELMy H-mode discharges in JET. Nucl. Fusion **51**, 103033 (2011)
22. Mantica, P., et al. A key to improved ion core confinement in the JET tokamak: Ion stiffness mitigation due to combined plasma rotation and low magnetic shear. Phys. Rev. Lett. **107**, 135004 (2011)
23. Mantica, P., et al.: Ion heat transport studies in JET. Plasma Phys. Control. Fusion **53**, 124033 (2011)
24. Luce, T.C., et al.: Experimental tests of stiffness in the electron and ion energy Transport in the DIII-D Tokamak. In: Proceedings of 24-th fusion energy conference, San-Diego, Rep. EX/P3-18 (2012)
25. Idomura, Y., Urano, H., Aiba, N., Tokuda, S.: Study of ion turbulent transport and profile formations using global gyrokinetic full-f Vlasov simulation. Nucl. Fusion **49**, 065029 (2009)

Concluding Remarks

This monograph describes in detail the development of the basic idea that transport in tokamak plasmas is dual in nature. The profiles of the plasma parameters and the fluxes of energy and particles are *separated*. This is the real meaning of self-organization. The profiles are defined by the magnetic configuration of the plasma, but the fluxes mainly by turbulence and to a lesser extent by binary collisions. On one hand, this idea is supported by numerous experimental observations of the conservation of pressure and temperature profiles. Some examples are included into the Introduction. On the other hand, the dual nature of transport is unexpectedly confirmed by the structure of the scaling law for the energy confinement time. Let us consider for example the scaling law ITER-98(y, 2) (4.30) established from a careful analysis of approximately 1500 discharges from 15 devices

$$\tau_E \propto I^{0.93} B_t^{0.15} \propto B_p^{0.93} B_t^{0.15}. \tag{1}$$

It is clear that the poloidal and toroidal magnetic fields (B_p and B_t) have a totally different influence on transport because the magnitude of the exponents differs by a factor of six. This fact cannot be explained without taking into account the effect of self-organization: for large values of the safety factor $q \propto B_t / B_p$ the pressure profile is narrow and the energy content is low. An increase in the toroidal field B_t will further increase q and thus enhances the negative effect. The other factor that determines confinement is the energy and particle transfer, since the increasing of B_p and/or B_t reduces transfer. In scaling (1) both factors determining confinement are mixed. Therefore the exponent of the poloidal field B_p describes the summary effect, as with increasing B_p the pressure profile becomes broader and the energy transfer is diminished. In contrast, the exponent of the toroidal field in (1) describes the difference effect, because with increasing B_t the pressure profile becomes narrower, but the energy transfer is still diminished.

In this monograph, the description of such a dual nature of transport is carried out in two stages. First, the problem of determining the canonical profiles (the effect of self-organization) is solved. Then, we develop a transport model that combines two aspects of the transport process: the self-organization and the transfer of energy and particles.

Yu.N. Dnestrovskij, *Self-Organization of Hot Plasmas,* DOI 10.1007/978-3-319-06802-2,

Of course, the proposed model includes a large number of phenomenological and empirical elements, and a more consistent theory needs to be developed. One important idea for such a theory, as it now seems to author, has been proposed in the work of Y. Idomura et al. (Chap. 6 [25]). Here (as it follows from the calculations by gyro-kinetic code) the transport in the ion channel has just a dual structure: over to some stiff ion temperature profile the avalanches of the energy and particles are rolled down with a period less than two orders of magnitude of the energy confinement time. Unfortunately, the further development of these ideas is unknown to the author. There remains, however, a hope that someone picks it up.

The manufacturer's authorised representative in the EU is Springer Nature Customer Service Centre GmbH, Europaplatz 3, 69115 Heidelberg, Germany. If you have any concerns regarding our products, please contact ProductSafety@springernature.com

Printed and bound by CPI Group (UK) Ltd, Croydon, CR0 4YY
15/07/2026
02167654-0003